ARE WE BEING
WATCHED?

ARE WE BEING WATCHED ?

The Search for Life in the Cosmos

Paul Murdin

with 36 illustrations, 29 in colour

Thames & Hudson

Acknowledgments

I should like to thank Professor Charles
Cockell, of the School of Physics and
Astronomy at the University of Edinburgh,
and Dr Leila Battison, of the Department of
Earth Sciences at the University of Oxford, for
their invaluable suggestions during my work
on this book.

Paul Murdin

On the jacket: iStockphoto.com/loops7

First published in the United Kingdom in 2013
by Thames & Hudson Ltd, 181A High Holborn,
London WC1V 7QX

Are We Being Watched? The Search for Life in the Cosmos
Copyright © 2013 Thames & Hudson Ltd, London
Text copyright © 2013 Paul Murdin

British Library Cataloguing-in-Publication Data
A catalogue record for this book is available from the
British Library

ISBN 978-0-500-51671-3

Printed and bound in China by Everbest Printing Co Ltd

To find out about all our publications, please visit
www.thamesandhudson.com.
There you can subscribe to our e-newsletter, browse or
download our current catalogue, and buy any titles that
are in print.

Contents

INTRODUCTION

In Search of New Lands and Peoples

To explore is an innate part of being human. It is a characteristic of all ages of people, from childhood onwards, and a feature of all ages of history. About 100,000 years ago the first modern humans set off out of Africa. They migrated from Ethiopia across the Arabian Peninsula into Asia and Europe, with their descendants eventually moving east and south into the Americas and Australasia. In this way human beings explored the world before they settled down to occupy the parts that they found most favourable.

Much later, during the late fifteenth and sixteenth centuries, Europeans embarked upon an age of exploration. They flooded out of their home continent to the far reaches of the Earth, driven by their imaginations and by the prospect of conquest and colonization. Christopher Columbus (1451–1506) and his crews set out in 1492 to sail west across the Atlantic Ocean; they found Central America, claiming possession for Spain. Not long afterwards Ferdinand Magellan (c. 1480–1521) circumnavigated the world, having searched for his route west by sailing down the eastern coast of South America, claiming it for Portugal. Spanish, French and British explorers vied with each other to explore and claim North America. A large area of the geographical globe diametrically opposite Europe remained blank, but there had been a continent hypothesized there as Terra Australis Incognita ('the unknown land of the south'). Different parts of it were discovered by Dutch and, later, British explorers in the next 200 years. By the late eighteenth and nineteenth centuries, the disposition of the continents of the Earth had been established; meanwhile, exploration continued, for example in the voyages of the Spanish ship *Pizarro* and the British naval ship *Beagle* to explore South America and the rest of the world. Such ventures mapped the land in detail, and surveyed its natural history. The data that they yielded gave rise to such scientific advances as Alexander von Humboldt's (1769–1859) discoveries in geography and meteorology, and Charles Darwin's (1809–1882) theory of evolution.

As late as the early years of the twentieth century, it was still possible to imagine isolated islands or plateaux where no one had ever set foot, just as Arthur Conan Doyle (1859–1930) imagined a 'lost world' in South America in 1912. Even today there remain some areas within the oceans that have never been seen by human eyes. But our exploration of the land areas of our world is now essentially complete.

In history, the psychological drive to explore may have been rooted in the human character, but in the age of exploration it was allied with the motivation to seek out material advantage: new land to develop and new resources to exploit. The expeditions were usually financed by capital from sponsors, often from the state treasury of the king or queen. As part of the exploration process, scientists were included as auxiliary crew, to give advice. For example, in 1585 the cartographer and polymath Thomas Harriot (1560–1621) was taken along on Walter Raleigh's (c. 1554–1618) expedition to Roanoke Island in the Carolinas; the astronomer Charles Green (1734–1771) and botanist Joseph Banks (1743–1820) were taken to Tahiti and Australia by James Cook (1728–1779) in 1768–71; the naturalist Alexander von Humboldt was taken to South America in 1799–1804; and the geologist and naturalist Charles Darwin in the 1830s accompanied the second voyage of the *Beagle*.

The new frontier of exploration is in space. The space age began in 1957 with the launch of the Russian Sputnik, a simple technology-demonstrator for the principles of spaceflight, and for the military capability of the USSR. The very next year saw the first successful American satellites, Explorer I and III (Explorer II failed to reach orbit), launched for the scientific investigation of the near-space environment of the Earth, and to show the competitive position of the USA's space programme. These two satellites discovered that energetic, electrically charged solar particles are trapped in the magnetic field of Earth, the so-called Van Allen Belts. This was the first major discovery of space-age exploration.

Human exploration of space has reached beyond near-Earth orbit only as far as the Moon, with the first manned lunar landing by Apollo 11 in 1969. All the larger planets in our Solar System have now been visited by robotic spacecraft, out as far as Neptune; the New Horizons space mission to Pluto was launched in 2006 and will in 2015 reach what was regarded for most of the twentieth century as the outermost planet at the edge of the Solar System (it is now regarded as the innermost dwarf planet of the Kuiper Belt: see p. 57). Video and digital cameras, using charge-coupled devices (CCDs) as detectors, have been our proxy eyes, and have viewed all the different kinds of bodies in the Solar System, including a good sample of the larger ones and a few of the smaller, sometimes at low resolution and from a great distance, and therefore rather indistinctly.

In the age of space exploration, scientists are often leaders, not followers. They too are supported by state finance; at least in part this support is altruistic, for the sake of the scientific knowledge that can be gained. Although there is no immediate sign of material profit in space exploration, no doubt this support is given with an eye towards resources that might be discovered and exploited in future. One resource that is in prospect is the discovery of extraterrestrial life.

Columbus, and the many other European adventurers who journeyed to the Americas, brought back to Europe stories not only of the peoples who lived there but also of previously unknown life forms. Tobacco, tomatoes and potatoes are well-known examples that have continued economic value. Darwin's finches in the Galapagos Islands are examples of formerly undiscovered life forms that had no direct economic value but which were key instances of species that changed the way we think, in this case about natural selection, environmental adaptation and the development of species. The new lands and the new biology turned out to have enormous social and economic value for the European countries that sponsored the exploratory voyages. The discoveries also expanded intellectual horizons as well as the literal ones of the sea voyages. Similarly, the discovery of extraterrestrial life will alter our perspective on our own lives, and provide raw material (tangible and intangible) for exploitation by biology and biotechnology.

Those in the twenty-first century who search for life beyond our planet Earth thus have something in common with Columbus, but with one difference: when he set out, Columbus was not sure what land lay beyond his horizon, but it was unthinkable that there was no life on it. We know what land there is in space, at least in the Solar System and in hundreds of planets orbiting the nearer stars, but nobody yet knows whether life does exist elsewhere in the Cosmos. Nevertheless, like Columbus we have a combination of facts and informed speculation to assist us in our search.

Our first voyages of discovery should certainly be to planets and moons that are not too distant from Earth, where we have reason to believe that life once existed or perhaps still exists, hidden away from hostile forces. The world's space agencies are already cooperating and competing to go to the nearer planets: the places in the Solar System where it is most likely that life could have evolved. In the future, our search could take us far beyond the bounds of the Solar System. If life does exist beyond our planet, it may be a long way from here, in some remote part of our Galaxy. The further away we search, the more space we will have to cover, and the more likely we are to find what we are looking for, but the more difficult that will be. Since there are literally billions[1] of distant planets, we must devise strategies to focus our search. So we should start by considering how we would look for life and the technologies we can deploy. After all, if we lack the means to conduct a search, we cannot possibly hope to discover new peoples.

The Cosmos is, of course, a very big place, and before we embark upon any attempt at discovery, like Columbus we should gather what knowledge we already have to guide us. For example, from what we know of our Solar System, can we gain some idea of where to look and what to look for? We can, as we shall see, be pretty sure of what conditions made life possible on Earth, although that by itself does not guarantee that life exists elsewhere in similar conditions, or that it does not exist where conditions are different.

The one certainty about life in the Cosmos is that it exists, in considerable abundance, on our own planet. If we examine the origins of Earth, the origins of life on it, the essential chemistry of living creatures and the conditions that make life possible here, can we develop search strategies that are likely to lead us to the discovery of forms of life elsewhere? Earth is, we will discover, a rather benign environment for life. But it also harbours life-forms that exist in very hostile places indeed. In fact, life may have originated in some of these regions. This suggests that, in our search for living creatures, we should scour even those planets that suffer extreme conditions.

Another way to refine our search would be to look for planets with the right kind of energy sources, atmosphere and environments hospitable to life; or at least climates that are not too inhospitable. Satellites, telescopes and probes provide us with plenty of evidence to help narrow our search to planets that can sustain life.

Perhaps the most intriguing question is whether any life-forms we encounter will be like us or very different indeed. Will alien life be a green slime, or a super-intelligent and advanced species of people to whom humans will appear very primitive? Of course, we will not know until we encounter some, but if we examine the experience of life on Earth we can generate plausible expectations.

This, of course, raises the question of how life emerges and develops. If life was brewed from a mixture of organic chemicals, where did those chemicals come from? What sparked them into life? Is life created locally, on Earth for example, using only ingredients on Earth? Or could it be that interstellar bodies have transported the ingredients of life across huge distances to deposit them on our planet?

If we are unlucky and fail in our search for extraterrestrial life, we might, nevertheless, discover a planet where life once existed but was made extinct for some reason. We have plenty of evidence to suggest that extinctions are not uncommon. Changes in climate, or large-scale natural disasters, may wipe out individual species or even all traces of life. We could still look on such planets for the residual traces of defunct life-forms. What forms might these take?

In this book, like a detective trying to solve a mystery with forensic science, I will examine many clues, big and small. I am an astronomer but I will draw on the skills of my scientific colleagues in many disciplines. I will try to interpret ambiguous and elusive evidence, in an attempt to discover whether extraterrestrial life exists, even though we do not know about it and remain isolated for the time being. Above all, I will try to answer the big question: is there anybody out there waiting for us to discover them, indeed encouraging us to do so? Or are we more than lonely?... Are we really alone?

1

The Century of Astrobiology

The twenty-first century is the century of astrobiology: this is the era in which we will discover life on other worlds, and learn from it. This will be a momentous discovery.

Astrobiology is the science of the nature and distribution of life in the Universe. As I write, astrobiology is a science about something that might exist, not something that exists for certain. I can make my hopeful prediction about the discovery of life in the Universe because in the last years of the twentieth century scientists have several times come so near to such a discovery without quite nailing it. There has always been a 'but' that has turned the initial thrill of the possibility of discovery into an agony of doubt, still unresolved. Radio astronomers have intercepted a signal with all the signs that it was both celestial and artificial, but it never repeated and remains an undeciphered mystery. Meteoriticists have found a meteor from the planet Mars containing fossil shapes similar to bacteria and biologically produced minerals, also apparently from Mars, although there were other terrestrial (and/or mundane) explanations. Space scientists have landed robotic spacecraft on Mars with experiments to test the soil for life and found the expected chemical activity, but, as I will explain in detail in chapter 13, it was a bit peculiar and there were no dead bodies in the soil. Astronomers have found hundreds of planets orbiting other stars, but their equipment has not until recently been capable of finding any earth-sized planets, and still has not found any that are truly Earth-like, although, as I will show later, it is likely that they are abundant. Space scientists have found niche habitats on remote planets in our own Solar System that are environmentally similar to habitats on Earth where life teems in abundance, but have not yet visited them to prove they are fertile.

Plans are being made. In the last years of the twentieth century, man was watching these worlds keenly and closely, scrutinizing and studying them as narrowly as a man with a microscope might scrutinize the transient creatures that swarm and multiply in a drop of water. Some may think of these worlds only to dismiss the idea of life upon them as impossible or improbable. At most, some men fancy there might be life upon Mars, inferior to themselves and ready to welcome a missionary enterprise. Yet across the gulf of space, our scientists' intellects, vast and cool and unsympathetic, regard these worlds with curious eyes, and slowly and surely draw their plans against them.

In the rather stilted nineteenth-century words of the above paragraph I have turned around the opening lines of H. G. Wells's (1866–1946) *War of the Worlds* (see pp. 174–75), which describe how we might have been seen by Martians plotting to invade Earth, to convey how, almost exactly one hundred years later, we are regarding those other worlds on which we might find life. We too plan to invade, after discovering evidence of life there. Scientists do not, however, have the killer cluster of evidence that life exists anywhere in the Universe to make it more than speculation that we would find it if we left our planet.

The reason why we must continue to try is that we can be sure that the life that we find in other worlds will be both like and unlike ours. It will be like ours because it looks as if there is one good way to make life from chemicals that are everywhere, in environments that are common, through processes that are natural: probably life somewhere else will have found the same way to come into existence. In any case, if the life we find is not sufficiently like ours, we will not recognize it as life at all.

But extraterrestrial life will not be the same as life here on Earth. The species that we find will certainly be different; biologists still find distinct and previously unknown species when they investigate unexplored environments on Earth, so we can certainly expect to find new species on other planets. Although they will probably be carbon-based, they may not function chemically in precisely the same way that our life does. Life here on Earth exists with the distinct biochemistry that it does because this was the way that life evolved on Earth. Some possible chemical architectures never happened, others may have started and never took off, and others may have started and evolved but, due to some chance event, died out. Life elsewhere could well have evolved in a biologically different way, and therefore with different capabilities. It would be a rich resource to know what those capabilities are, and how extraterrestrial life differs from ours in its make-up and machinery.

So, if we find new forms of life, we will learn from the discovery, by comparing and contrasting them with ours. We will extend our knowledge of biology and biochemistry, and, as a result, make advances in medicine and biotechnology. Who knows where that will lead? We see vaunted in science fiction (for example, in such films as *Blade Runner* and *Alien*) the value of off-world mineral deposits and other natural resources. The intellectual capital found from the examination of extraterrestrial life is likely to be socially and financially much more valuable: the greater understanding of life that we will acquire will lead to new medicines, new chemistry, new industrial processes. But of course we first have to find extraterrestrial life in order to acquire the new knowledge.

There will, however, be other effects upon our way of looking at things, of understanding our position in the Universe, which will be more momentous. The discovery 400 years ago that our planet was not the centre of the Universe but was one typical planet of the half-dozen planets then known in orbit around the Sun (Mercury, Venus, Earth, Mars, Jupiter and Saturn) changed our world picture for ever. The same will surely be so of the discovery of life elsewhere in the Universe, even if we cannot say precisely what the impact will be.

According to anthropologist Kathryn Denning, the search for extraterrestrial life might be like the hunt for the unicorn, the story woven into a tapestry from Brussels (*c.* 1500) that now hangs in the Metropolitan Museum of Art in New York.

> A noble visionary few go in search of something rare and beautiful that may not even exist; they find it; it resists capture; there is a drama of some sort (in this case the death and resurrection of the beast); they deliver it to the ruler who rather fancies rare and beautiful things that no one else has; and the beast lives happy ever after in blissful contentment in captivity, utterly tamed.

She goes on to warn, however, that:

> even if we can hunt life and then triumphantly capture it, and keep it in a cage, domesticate it...beware of thinking that we can do that with life's meaning as well. The idea that we can predict what a detection of other life will mean to the world, that we can use our cleverness to control what that information will do in the world, that we call those shots...this is the stuff of fairy-tales.

We can turn to writers of modern fairy-tales for some ideas. Science-fiction writers invite us to fear hostile colonists, or to welcome wise and peaceful gurus. They imagine parasitic, egg-laying arthropods, or time-travelling mind-readers. Science itself suggests either that the reality will be more modest and, face to face with our first extraterrestrial, we may be confronting a bacterium;

or that on some principle of universality we may be able to talk to someone who looks and thinks pretty much like ourselves, a bipedal, thinking, warm-blooded being like ET in Steven Spielberg's film of that name. What will our reaction be? We may want to plumb the mind of our ET for new technology, wisdom or just a different point of view. We may erect defences against whatever change is threatened by the contact. We may want to convert ET to our own ways or colonize ET's terrain, real or virtual.

The prediction that I made in the opening paragraphs of this book could be false and optimistic, and grossly overstate the possibilities. Maybe we will not ever find extraterrestrial life of any kind; maybe we will not ever contact intelligent extraterrestrial life. Suppose our efforts are fruitless and we conclude that we are quite alone, with no one else in the Cosmos to talk to. Will we accept this rationally and with resignation, as part of being human, as we accept our own mortality? Will we despair and collectively go off the rails? In the modern world picture we contrast ourselves in scale and timespan with the vastness and duration of stars, galaxies and the Universe, and feel humble. If it turns out that we are unique, will we change our minds about how unimportant we are? Perhaps we will return to the medieval belief that we are the focus of the Universe's attention, with 'dominion over the fish of the sea, and over the fowl of the air, and over the cattle, and over all the Earth'; if he had been alive now, would the writer of Genesis[1] add 'and over all the Universe'? This would not be a healthy attitude for mankind to readopt: in Greek tragedy, after hubris, there always follows nemesis.

But these possibilities are in the future. What is in the present, and what raises the subject of this book above mere speculation, is the scientific search for alien cultures, other worlds and life in the Cosmos.

2

Other Worlds

How and where did life originate here on Earth? Or perhaps it began in some faraway interstellar location? Are there other planets where life exists now? What kind of life might there be elsewhere? Will it be very primitive, or more advanced than us? Are we being visited at the present time by aliens? Can we talk to them? Are we the descendants of extraterrestrials or they of us? What makes life possible?

When, perhaps at a party, people learn that I am an astronomer, the talk often turns to such questions as these. Contemplating life in the Universe seems to be a fundamental human concern, as it is one that touches upon our very origins. The answers to 'where did I come from?' or 'who am I like?' may differ - according to whether such questions are asked by a child of its parents, or by the parent in a library, church or observatory - but the questions are universal.

That we all ponder the same questions is perhaps because the answers are so awe-inspiring. The implications are momentous, and span two extremes: first, that as intelligent life, we are alone; and second, that we are just one species among countless kinds of life on planets everywhere in the Universe.

At the one extreme, if our world is the sole place where life flourishes, we may be frightened of the implication that the Universe echoes with sterile emptiness. Blaise Pascal (1623–1662), the French philosopher, expresses this horror in his book *Pensées* (*Thoughts*), which is a rather pessimistic defence of Christianity, but also a hopeful analysis of what it means to be human: 'The eternal silence of these infinite spaces strikes me with terror.' Alternatively, we might react with a feeling of wonder at the responsibility of being unique, and the possibility that we are the sole focus of some grand purpose, and gain an inflated sense of our own importance.

At the other extreme, if life is everywhere, we might be surrounded by potential companions spread far and wide across the Universe, just as on Earth we have neighbours on our street, compatriots, and a global social network, distributed on the internet. We might be comforted by the thought of friendship or support. Or perhaps we might be frightened, certainly impressed, by the idea that other occupants of the Universe may be more potent than we are, possibly visiting us even now in flying saucers. We might fear that they are implicitly threatening us with terrible weapons; or we might hope that they will be able to impart to us their wisdom and advanced knowledge.

* * *

As a first step to deciding whether there is anybody else residing in the neighbourhood of our world, we might start by asking if there are any other worlds beyond our own, the assumption being that life needs one to live on. If there are many worlds in the Universe, then life could in principle be present on any of them, but if our own is unique, then life probably is too; what is here would be all the life there is.

The philosophical dispute as to whether the life on our world is alone in the Universe is an ancient one, going under the name of the question of the 'plurality of worlds'. I trace the emergence of the conflicting points of view in Chapter 3. In Western philosophy it dates back to the earliest Greek philosophers. The Creationist movement harks back even earlier than that, to the authority of the writers of the Book of Genesis in the Old Testament. Still today, the proponents of Creationism go so far as to set aside a range of scientific evidence from astronomy, geology and biology; claim an unfeasibly short time for the age of the Earth; deny the theory of evolution; and focus on Earth as the unique abode for life.

Others have set out a much more difficult, uncertain, incomplete, but evidence-based approach to the exploration of life in the Universe. Nowadays investigations happen in institutes of geology, astronomy and astrobiology, across the world, but the modern understanding of the nature of the Solar System first began in the sixteenth and seventeenth centuries. At this time, the Sun, Moon, Mercury, Venus, Mars, Jupiter and Saturn – all of which had previously been classified together as 'planets' – were separated into three categories. The Sun was understood to be a star, the Moon a satellite of Earth, and the rest were grouped together as planets, our Earth included. There was no need for further philosophical speculation: empirical science had shown that there were at least five more worlds than our own.

But this of course was not the end of the search. If the Sun is a star and has a planetary system, then other stars may have planetary systems too. Until the end of the twentieth century, this remained a reasoned, but unproven, guess.

There is little that is more difficult for astronomers to detect than a planet that is in orbit around another star. Because stars are a very long way away, a planet would appear to us to be very close to its parent star. Imagine a pea and a pinhead separated by the diameter of a coin, and viewed from across the Atlantic Ocean. This represents in scale the separation between a planet and its star for one of the nearest planetary systems. The problem of detecting the planet by recording its image is compounded by the fact that a planet emits or reflects very little light compared to a star. Using a camera to identify a planet, such as Jupiter, in orbit around the nearest star is a million times more difficult than distinguishing a glow worm next to a car headlamp, as seen across an ocean. This explains why the first indications of extra-solar planets (or 'exoplanets': planets in a planetary system orbiting a star other than our Sun) were indirect.

When the breakthrough came, at the end of the twentieth century, the dam cracked and then burst: at first a trickle of ones and twos of planets, then a flood of hundreds, and now thousands. There are indeed other worlds orbiting other stars. Not all stars have planets, but some stars have many. Overall, there are probably as many planets in orbit around other stars as there are other stars.

* * *

The very first exoplanetary system was found in 1992, when Aleksander Wolszczan (b. 1946), a Polish-born American radio astronomer, was timing the rapidly rotating pulsar PSR1257+12 (at a distance of about 1,000 light years, in the constellation Virgo). He noticed that its pulses alternately arrived earlier and later than expected, in what proved to be a distinctive triple rhythm. Wolszczan's interpretation was that the pulsar was being pulled nearer to and further from Earth by three planets in orbit around it. The three planets range from moon-sized to a few times Earth-sized.[1]

Wolszczan's planetary system exists in circumstances completely different from our own: its star is the remains of a supernova explosion, and its planets are a second-generation planetary system, formed not at the birth of a star, as were the planets of our Solar System, but at its recent demise. The system's central star is a neutron star, a star as unlike the Sun as it is possible to get. It bathes its planets in a lethal flux of radioactive particles; the planets are very unlikely to be an abode of life. Indeed, it is hard to imagine that if there used to be life here it survived the supernova explosion. There would probably not have been time since then for new life to have evolved on the presumably sterile planets formed from the supernova debris. Most likely Wolszczan's system is, and always was, completely dead.

* * *

Then astronomers found the first planetary systems orbiting stars that were much more like our own Sun. The breakthrough came in 1995. In April 1994 two astronomers from the University of Geneva, Michel Mayor (b. 1942) and Didier Queloz (b. 1966), began a programme to find Jupiter-sized planets in orbit around solar-type stars. They did this by looking for changes in the speed of the stars that could be due to the gravitational pull of planets.

Mayor and Queloz designed their search strategy from analogy with our own Solar System. It is only an approximation to say that the planets of our Solar System orbit around the Sun. It is more accurate to say that the Sun and the planets orbit around their common centre of mass, although, because the Sun is so much more massive than any planets, the centre of mass of the Solar System is actually inside the Sun. As a result, the Sun scarcely moves in its orbit around the centre of mass. Its orbital speed is 13 m per second. That is not much in cosmic terms – it is scarcely faster than an athlete can sprint. The delicate quiver of the Sun is largely due to the gravitational pull of the most massive planet in the Solar System, Jupiter. The main period of the Sun's motion is therefore the same as the orbital period of Jupiter: twelve years.

When Mayor and Queloz started their programme, no existing instrument was capable of detecting this 'quivering' motion in stars other than the Sun. They developed a very sensitive instrument designed for the job. They also negotiated long periods of access to what, in modern terms, is a very modest telescope. This enabled them to carry out frequent observations in order to monitor the changes of speed that they were seeking. They drew up a list of 142 Sun-like stars. They chose stars that were nearby and therefore bright, with plenty of light to analyse, to suit the moderate light-gathering power of the smallish telescope. Another selection criterion was that the stars did not show, at coarser accuracy, any large changes of speed that would suggest they were members of double-star systems. Astronomers calculate that a planet in a double-star system would loop in complicated orbits between the two stars, and in a relatively short time could be ejected from the system altogether, unless the planet orbits at large distance around two stars so close that effectively they seem like one. (This is the case with the example of a planet in a double-star system that was discovered at the end of 2011. It is informally known to astronomers as Tatooine, the name of the fictional planet in the *Star Wars* movies that has two suns.)

Additionally, some stars have a very active surface, with surging motions of their atmospheric gases, which produce effects that can look like changes of speed. Known examples of similar stars were eliminated from the list. (There could be planets in orbit around such stars, although, for the purposes of this book, they would not be interesting – the activity of their sun is likely to sterilize them and they would not sustain extraterrestrial life. But evidence for the planets would be masked, so astronomers are reluctant to look.) The possibility that some examples of active stars remained in their sample injected a note of caution into Mayor and Queloz's observing programme – they would have to be sure that changes of speed that they saw were truly due to an orbit. This meant that they expected to have to plod on in the monitoring programme for a long time, something like the twelve years of Jupiter's orbit, or perhaps twice that. Only by completing observations of more than one cycle of the oscillations could they be sure that the motion was truly periodic and that the changes had the right characteristics.

Mayor and Queloz were therefore very surprised when, within only eighteen months, they had discovered their first planet, orbiting the star 51 Pegasi, 45 light years distant. The star's oscillation was much greater than expected, and its period was very much shorter: 4.2293 days, much, much less than 12 years. The planet, 51 Pegasi b, has roughly the same mass as Jupiter.[2] Its short orbital period indicates that the planet lies much closer to its sun than Jupiter does to our Sun – in fact, closer than any planet in our Solar System. The distance of 51 Pegasi b from its parent star is only one-twentieth of the Earth–Sun distance. The planet Mercury, closest to the Sun, has a period of 88 days and its distance from the Sun is just under half the Earth–Sun distance.

The discovery was leaked at a conference in 1995 and quickly spread around the world on the astronomers' rumour machine. It caused a public sensation – the first planet discovered outside the Solar System. Of course, extraordinary discoveries need strong proof, but that quickly came from a team in the US. They had already organized a brief four-day run in 1995 with the powerful Lick Observatory telescope, just long enough to encompass one orbit of 51 Pegasi b, if the weather was clear. The team was led by Geoffrey Marcy (b. 1954) of San Francisco State University and Paul Butler (b. 1963) of the University of California, with members from the High Altitude Observatory and the Harvard Smithsonian Center for Astrophysics. Marcy and Butler had been carrying out a programme exactly like that of their European colleagues and competitors Mayor and Queloz. The star 51 Pegasi was not on their original observing list. After hearing of the Swiss discovery, the Americans altered their expectations about the orbital periods of exoplanets from decades to days. They quickly confirmed the existence of 51 Pegasi b. They went on immediately to find further examples of extra-solar planets, some of them in archives of their previous observations. They had not yet examined their data closely because they thought there was no rush, and so were pipped to the post in the race for exoplanets.

*   *   *

The technique used by Mayor and Queloz and by Marcy, Butler and their colleagues works best for massive planets, such as Jupiter, because these have a bigger gravitational pull and produce a greater reflexive motion in their sun. Another technique that has been successfully used to discover exoplanets is to look for a 'winking star': a star the light of which is periodically dimmed a little (1 per cent or less) by the transit of a planet across its face. This has the potential to detect planets of Earth's size. There are several projects (two of them in space, called COROT and Kepler) in which telescopes stare unceasingly at a field of many stars, recording their images and measuring the brightness of each one, in an attempt to see these periodic winks. Such projects represent a considerable data-processing task; there may be millions of stars in the field of view of the telescope, each of them recorded every few minutes. Each has to be looked at to see if it is constant or whether its brightness changes in the way expected from the transit of a planet across its face.

The Kepler space project uses automated computer techniques to scan the data for the winks. The computer finds most of them, but there are some cases in which the data is in some way unusual – the 'wink' may appear towards the end of a sequence of measurements, for example – and the computer is not intelligent enough to cope with the problem. The Kepler space project feeds all its data on to a website where it can be examined by eye by anyone, for a second look. A team of so-called 'citizen scientists' called PlanetHunters – volunteers, members of the public – has sprung up, taking on the task of looking through the data, fired up by the possibility of discovering another planet. By spring 2012, the PlanetHunters had searched 11 million observations and found 34 confirmed and potential exoplanets.

When the transits are detected, the next thing is to study the star to see whether it wobbles or quivers in the right way. This is reckoned to be the clinching observation that proves that the variability of the star's brightness is due to the presence of a transiting planet.

All in all, between the mid-1990s and May 2012, nearly 800 exoplanets were found and confirmed. Due to technical constraints, however, only very specific kinds of exoplanets have been found thus far, representing a minuscule proportion of the probable total. Most of the stars that are currently known to have exoplanetary systems are near our Solar System: in a volume of space up to about 1,000 light years away (our Galaxy of stars is 100 times this size, 1 million times its volume). Because the stars with which exoplanets have been found are so close, they are therefore relatively bright; a lot of light is required to measure accurately the small radial velocity shifts produced by jupiters or the small winks produced by earths. Nearly all the exoplanetary systems discovered consist of a single, large, Jupiter-sized planet orbiting close to a central star, because these larger planets are easiest to find. They are gas giants and have no stable surface. It is possible that they have rocky satellites on which life may be found, but there is probably no life on the planets themselves.

In February 2009, however, the French space satellite COROT, a forerunner to the Kepler project, located a planet, COROT-Exo-7b,[3] which is only 1.7 times the diameter of Earth, and perhaps 5 times its mass. With an orbital period of only 20 hours, the planet is very close to its sun, and its surface has a temperature of more than 1,000°C. It must have a lava-like surface, and is not a very promising place to expect to find extraterrestrial life. The same is true of the planet known as Kepler-10b,[4] which was discovered by the Kepler space mission. Kepler-10b has a diameter 1.4 times, and a mass estimated at up to 4.5 times, that of Earth. Its density is about 9 g per cubic centimetre: comparable to iron, and considerably denser than Earth. It is much closer to its star than the planet Mercury is to the Sun, and the temperature of its surface is 1,500°C, hot enough to melt iron. Finally, the Kepler spacecraft has discovered a planetary system of five planets orbiting a star catalogued as Kepler 20, of which the two smallest are between 0.4 and 1.7 times and 0.7 and 3.0 times the mass of the Earth. The temperatures of their surfaces are 760°C and 427°C respectively. Again, although all these planets are in some critical ways like Earth, they too are improbable places to find extraterrestrial life.

By the end of 2011, the Kepler mission had detected more than two thousand possible planets, although they need confirmation in order to eliminate events in the data that look similar but are not due to the presence of planets. Two hundred are Earth-sized and fifty may be in the 'habitable zone' of their planetary system. This is the range of orbits that position the planet at just the right distance from its sun to have a life-friendly temperature (chapter 4). This number includes six likely planets that seem to be not much bigger than Earth. A small object, known as Gliese 581 g, is an unconfirmed exoplanet, orbiting the red dwarf star Gliese 581.[5] A red dwarf is a star much like the Sun, but smaller, and therefore cooler and redder. There are lots of these stars, and it is important to characterize the planetary systems that red dwarfs have, because they are likely to be the most typical planetary systems in the Galaxy. Gliese 581 g is one of a retinue of six planets, labelled a to g, in orbit around the star.

They were all discovered by looking at the 'wobble' of the parent star. Gliese 581 g is about four times the mass of Earth. As I write, the existence of Gliese 581 g is unconfirmed: some astronomers think that it might be a spurious effect of one set of data. If it really exists, Gliese 581 g is in its star's habitable zone and it may be a rocky terrestrial planet like ours: a good place to look for extraterrestrial life.

As a final example, the roughly Earth-sized planet Gliese 1214 b was discovered in 2009 by a ground-based set of robotic telescopes on Mount Hopkins, Arizona, which were looking for transiting exoplanets. Gliese 1214 b whizzes around its parent sun in not much more than a day, and its sun is a red dwarf, not very massive. All this has made it possible to calculate the size of the planet fairly accurately: it is 6.5 times the mass of Earth and 2.5 times its size. The planet, a so-called super-earth, has a low density. By calculating how a planet with these global characteristics could be made up, astronomers have deduced that it may consist of a rocky core surrounded by a deep ocean, although there is no direct evidence that it has any water – astronomers have not obtained a spectrum, for instance, that proves water is present on the planet. It is not very hot – although its average temperature is above 100°C – and it is likely to have an atmosphere of steam. Since it appears to have formed further out in its planetary system and migrated in towards its parent star, the whole planet, or at least its ocean (if it has one), is likely to be evaporating.

* * *

The planetary systems discovered so far show differences from our Solar System that make life on most of the examples that we know of difficult to sustain. Are they typical? Where else can we look? Does life evolve only on planets around solar-type stars? Are the exoplanetary systems that we know of stable; do they last long enough for life to evolve there? How can we look for life on these planets? I shall return to these questions in the following chapters of this book. But for the moment, I want to examine the question of how planets form. If there are so many planetary systems, they must be formed by a natural, common process. The principles of the process for the creation of our Solar System have been known for two centuries. If we can prove that this is how all planets form (save for some exceptional cases, such as Wolszczan's system), we can infer how common exoplanets are, and therefore how widespread life may be.

In 1796, a French astronomer, Pierre-Simon Laplace (1749–1827), proved mathematically that the shape of the Solar System, in which the planets all orbit the same way round the Sun in a flat disc, is the same as it was when it was formed: it has stayed that way since birth. He had found something that supported an idea put forward by the Swedish scientist Emanuel Swedenborg (1688–1772) in 1734 and the Prussian philosopher Immanuel Kant (1724–1804) in 1755. This is that the planets condensed out of a flat nebula – a 'cloud' of gas and dust – whirling around the Sun, an idea that goes under the name of the Nebular Hypothesis.

Let us follow the way that the Sun and Solar System began, from their early moments in a contracting gas cloud, as first calculated by the Japanese astrophysicist Chushiro Hayashi (1920–2010) in 1960. The protostar that became the Sun started as a collapsing sphere, but went through spasms in which it ejected a stellar wind in every direction and squirted jets of material from its poles. The stellar wind became flattened into a disc of gas clouds orbiting the star.

The protostar took up a rapid spin as it contracted with increasing speed in the interstellar cloud, just as, in a final flourish at the end of her dance routine, an ice skater will spin faster as she brings her arms closer to her body. The star slowed by throwing off material into the nebular disc of a protoplanetary system. This system was mostly hydrogen and helium, but it also contained elements that had been made in old stars and exploded into the interstellar cloud. Some of these elements made dust grains: grains of graphite (the material that makes the 'lead' of a pencil) and little diamonds, both made from carbon, and silicate grains (sand) made from silicon and oxygen. Each dust grain individually orbited its parent star. The grains were warmed by the parent star of the protoplanetary nebula and re-radiated their heat in the form of infrared radiation.

The first protoplanetary systems were found by homing in on this infrared radiation. The very first was found in 1966 in the Great Nebula in the constellation Orion by the astronomers Eric Becklin (b. 1940) and Gerry Neugebauer (b. 1932) from the California Institute of Technology (Caltech). Invisible to optical astronomers, the 'BN Object' (from the discoverers' surnames) is bright in the infrared. The InfraRed Astronomy Satellite, IRAS, discovered further protoplanetary systems in 1983, including dust discs orbiting the stars Zeta Leporis, Vega and Beta Pictoris. In the last star, planets orbit in and around the disc: astronomers were first able to infer that they are there from their effect on the dust grains. The planets' gravity perturbs the orbits of the individual dust grains and distorts the shape of the disc. In 2008/9, astronomers using the Very Large Telescope of the European Southern Observatory succeeded in imaging one of the planets of Beta Pictoris (1; bold numerals in parentheses relate to images on pp. 153–68. Captions to the images begin on p. 169.). It is a planet considerably larger than Jupiter, orbiting at a distance from its star comparable to the distance of Saturn from the Sun.

The first direct images of protoplanetary systems were made in 1992 by Robert O'Dell (b. 1937) of Rice University with the Hubble Space Telescope: dust discs silhouetted against the luminous background of the Orion Nebula, each with a central star, just as the Nebular Hypothesis had imagined (2). The Great Nebula in the constellation Orion is a dense cloud (*nebula* is Latin for 'cloud') of gas (3). It is one of the nearest nebulae to us and is therefore a well-studied example. The gas of the nebula is energized by the stars that are in it and shines brightly, silhouetting anything that lies in the foreground. (This is a dangerous environment for protoplanetary discs to be born in. They can be distorted by the flows of hot gas from the nebula and their planets aborted. See 4.) The stars in the Orion Nebula are all young, and it is often referred to as a stellar nursery. In fact, as well as some recently born stars, there are some that are still forming, including the objects that O'Dell had found. They were termed 'proplyds', a contraction of 'protoplanetary discs'.

In the protoplanetary nebula, grains of dust stuck together in solid lumps, which, being large, were then not easily destroyed or expelled. The lumps grew to kilometre sizes. (At this stage they are called 'planetesimals'.) The gravity of the planetesimals was high enough to cause them to attract one another; a large planetesimal attracted smaller ones, and grew bigger, so its gravity increased, so it grew bigger still, and so on. This process is called 'accretion'. It produced a number of protoplanets in orbit in our Solar System. Accretion stopped when the protoplanets had fed on everything around them. Material in the more distant part of the gas cloud was less affected by the heat of the star, so it became gas-rich and built up the gas-giant planets. The planet Jupiter grew the fastest and the most. Some of the small fragments that were left over from the protoplanetary nebula, and did not accrete onto any planet, fall to Earth from time to time. They are a particular kind of meteorite called a chondrite.

There are some possible situations in which this process for the formation of planets might be disturbed. Maybe there are hot, bright stars nearby that cause the protoplanetary nebula to dissipate before planets condense in it. But on the whole, in this picture, the formation of planets is a natural, integral part of the process of the formation of a star. This encourages the belief that planets are abundant in the Universe, since we know by simple inspection of the night sky (and even more so by inspection of well-exposed images from modern telescopes) that stars are very numerous. We know directly, from the number of planets that have been discovered, that at least 10 per cent of solar-type stars have jupiters, with the true number of planets being much larger because of the limitations of astronomers' techniques in detecting smaller bodies.

Estimates of the fraction of stars that have earths are still very imprecise. It may be as large as 40 per cent. There are billions of planets in our Galaxy, and the ancient question of the plurality of worlds has been definitively settled: there are many, many, many! But are any of them alive?

3

In Search of Alien Intelligence

If life does exist elsewhere in the Solar System, on neighbouring planets, there is some chance that we could find it by visiting its habitat. We already investigate places that are up to about ten years' travel time away; it would take only a matter of years to travel to most planets in the Solar System. Space agencies create protocols that hold together the complex procedures necessary to operate robotic spacecraft for such long periods of time. We have the political and legal frameworks for a space agency to remain functional for decades. At a laboratory level, the agencies are able to maintain handbooks and other physical and intellectual capital to link the decisions taken by the makers of the equipment with those people who are to operate it a decade or more later.

The Rosetta spacecraft, for example, was built in the 1990s and will land on its target comet in 2014. When it does so, it will be necessary to operate the lander through on-board computers that are programmed in a language that was already obsolete at the time of the launch in 2004. Duplicate operating systems and handbooks that fossilize the original state of the computers are locked in safes at the European Space Agency (ESA) and the manufacturing laboratories, so that it will remain possible for future controllers to land the spacecraft and even diagnose and correct faults.

As a society, human beings have been honing these social and organizational skills of continuity for centuries. A fine spirit, such as cognac, can be matured not just for years but for decades, and barrels of old brandy are kept under two locks in the cool, dimly lit vaults of family distillers. The task is passed from generation to generation, a pair of keys kept separately by two family members to ensure the integrity of the oldest and most costly spirits; if customers lost confidence in the continued preservation of the matured cognac, it would lose value. In this example, of course, the intention of the lapse of time is to improve what is to be preserved; in the case of a space experiment, the intention is to maintain it unaltered.

These organizational skills, and the necessary technological ones, make interplanetary travel possible; robotic spacecraft have been sent out to Neptune and even beyond. Voyagers 1 and 2 are the spacecraft that have travelled for the longest time and farthest distance in their explorations. Currently, the craft have been in space for thirty-six years since their launch by NASA in 1977, and have travelled more than four times the distance to the outermost planet of the Solar System, Neptune. Initially, Voyagers 1 and 2 were used to study the planets of the outer Solar System, but they are now going on through the heliopause, the boundary of the atmosphere of the Sun. After that we can say they have truly left the Solar System.

Human exploration is not yet demonstrated at such distances and amounts of time; we are, however, reaching the capability required to go to the nearer planets, including Mars (minimum return mission duration about eighteen months). A simulation of a Mars spaceflight was completed by the Russian and European space agencies in November 2011, when six 'astronauts' left a 'spaceship' – actually windowless living modules in a warehouse in Moscow – after 520 days locked inside, suffering the confinement and stress of interplanetary travel (but not the weightlessness, radiation and danger). The longest human interplanetary missions, the Apollo 15 and Apollo 17 journeys to the Moon in 1971 and 1972, were much shorter, at just over twelve days. Mission durations in near-Earth orbit have been more impressive. The longest manned spaceflight lasted more than a year: the Russian cosmonaut Valeri Polyakov (b. 1942) stayed at the Russian Mir Space Station for 437 days, starting on 8 January 1994. One astronaut, Sergei Krikalev (b. 1958), has spent more than two years (803 days) in space, in six separate flights. The Mir Space Station was occupied continuously (by a succession of astronauts) for ten years; the International Space Station surpassed Mir's record in 2010 and is likely to continue in occupation for more than twenty years.

It is thus conceivable that in the future, as human space travel progresses, we could visit another planet in our Solar System and come face to face with extraterrestrial life, be it bacterium, arthropod or vertebrate.

* * *

We might be able to contact life on the planets of the Solar System, but if there is life on a planet orbiting another star, what are the chances that we could meet it face to face? The nearest star is four light years away, which means that light takes four years to travel from it to us. A spacecraft, travelling more slowly, would take much longer: a hundred or a thousand times longer! To make a readable book or watchable film, science-fiction writers have had to invent ways to get round this: humans hibernate during the long trip (as in the film *Alien*), or the spaceship has a 'hyperspace drive' (or a transportation device, such as a 'teleporter') to land people on a planet instantly, without time-consuming distractions from the story. Somehow these devices, as seen in *Star Trek*, make it possible to move quicker than the speed of light.

Maybe this way of travelling faster than light will happen one day, but for us to visit another star, even the nearest, let alone any other star in the solar neighbourhood or beyond, is impossible with our current technology, or even with the technology of the foreseeable future. We can only dream of directly visiting a planet outside our Solar System to see if it holds life.

In the meantime, we can hope that life from planets around other stars will instead come to us. Of course, if there are beings that can do this, by possessing the technology for interstellar travel (a flying saucer or its equivalent), they must be not only intelligent life but also more advanced than us.

The alternative may be that there is no practical technology for interstellar travel. We might never meet intelligent extraterrestrial beings, whether on their planet or ours.

* * *

If, as yet or ever, we cannot physically visit a planet orbiting another star, could we nevertheless make contact with its inhabitants? Radio or light transmissions travel at the speed of light and we could communicate by those means on the same timescale as visiting another body in the Solar System. The dialogue would be stilted, with a gap of years between the question and the answer, but it would be possible.

Communications that we have already generated, without expectation of answers, reach out tens of light years. Regular radio transmissions began in 1920, and television a decade later, which means that there is a sphere of radio transmissions within a radius of about ninety light years around the Earth. Those broadcast earliest are on the outer skin of the sphere; for example, such old programmes as the racially stereotypical *Amos and Andy*, from the 1940s and 1950s. They cannot be recalled, so when these transmissions arrive at another planet we might wince at the first impressions they create of us. More recent radio transmissions, from radar systems and mobile phone networks, are more powerful, but lie within the central core of the sphere. These will eventually reach a greater number of stars at further distances, but as yet have not travelled that far because they started out later.

Conversely, of course, it is possible that extraterrestrial aliens are already communicating with us. More than a century ago, in 1899, the radio pioneer

Nikola Tesla (1856–1943), working in Colorado, picked up noise on his early radio receiver that he described as 'the pulse of the globe, as it were...every electrical change that occurred within a radius of eleven hundred miles'. He was struck by the presence of repetitive signals, which seemed to have some systematic pattern, among the random noise of lightning and the aurora. He later wrote,

> I can never forget the first sensations I experienced when it dawned upon me that I had observed something possibly of incalculable consequences to mankind.
>
> I felt as though I were present at the birth of a new knowledge or the revelation of a great truth. Even now, at times, I can vividly recall the incident, and see my apparatus as though it were actually before me. My first observations positively terrified me, as there was present in them something mysterious, not to say supernatural, and I was alone in my laboratory at night.... It was some time afterward when the thought flashed upon my mind that the disturbances I had observed might be due to an intelligent control. Although I could not decipher their meaning, it was impossible for me to think of them as having been entirely accidental. The feeling is constantly growing on me that I had been the first to hear the greeting of one planet to another.... At the present stage of progress [in 1901], there would be no insurmountable obstacle in constructing a machine capable of conveying a message to Mars, nor would there be any great difficulty in recording signals transmitted to us by the inhabitants of that planet, if they be skilled electricians. Communication once established, even in the simplest way, as by a mere interchange of numbers, the progress toward more intelligible communication would be rapid. Absolute certitude as to the receipt and interchange of messages would be reached as soon as we could respond with the number 'four,' say, in reply to the signal 'one, two, three.' The Martians, or the inhabitants of whatever planet had signalled to us, would understand at once that we had caught their message across the gulf of space and had sent back a response.

The tantalizing, repetitive signals that Tesla noticed have never really been explained. They were presumably spurious patterns generated at random in the noisy signal, but, even so, twenty years later, the Italian radio pioneer Guglielmo Marconi (1874–1937) also thought that he had picked up similar transmissions from outer space, possibly from Martians. In 1919, Marconi expressed hope that in the future it would be possible to communicate with intelligence from other stars: 'The beings there ought to have information for us of enormous value,' he said. In an editorial in January of that year, commenting on Marconi's suggestion, *The New York Times* warily suggested that we might in this way receive knowledge for which we are unprepared, and that we should 'Let the Stars Alone'.

The most modern radio telescopes can readily hear radio transmissions, not only from within the Solar System, but also from other nearby stars. The Square Kilometre Array (SKA), a planned radio telescope of 3,000 interconnected dishes adding up to the eponymous collecting area, is likely to be located in desert areas in South Africa and Western Australia, far away from terrestrial interference. It will be able to detect the extraterrestrial equivalent of a mobile phone system within fifty light years of Earth. When it is complete, the SKA will be able to identify such a system only if the extraterrestrials began broadcasting more than about fifty years ago, as the signals need time to reach us.

From 1960, systematized (but still fruitless) searches that listen for intelligent signals deliberately sent to us have been implemented under the name of SETI, the Search for Extraterrestrial Intelligence. SETI was first directed by radio astronomer Frank Drake (b. 1930), and is based on analysis by two young gamma-ray physicists at Cornell University, Philip Morrison (1915–2005) and Giuseppe Cocconi (1914–2008). In an article in the prestigious science journal *Nature*, in 1959, they speculated that there are large numbers of planets on which exist civilizations who have developed radio communication technology, and who may want to communicate by radio with other worlds, or even specifically with us. Is this fanciful, or could there be many such civilizations?

* * *

If asked about the possibility that extraterrestrial life exists, many people will start with the observation that there are lots of stars that are similar to the Sun (5). It is inconceivable, they might go on, that our planetary system is the only one. In fact, there are so many stars that there must be other inhabited planets with life on them, even intelligent life. This argument has been articulated for 2,000 years (see the discussion starting on p. 43). Modern science is in support, up to a point, and astronomers have taken this general feeling and tried to give it numerical substance, in a formula known as the 'Drake Equation'.

The formula is named after Frank Drake. In 1961, at a meeting on the subject of interstellar communication attended by representatives of various scientific disciplines, he tried to focus discussion and draw people together with an equation, the terms of which became the agenda for the meeting. This equation would calculate the number of communicating civilizations in the Galaxy – or at least make a plausible estimate – to see if there might be none, one, several, hundreds or millions of groups of beings with whom we might exchange messages.

Drake's equation started with the fundamental idea that the number of communicating civilizations that exist at any one time is the rate at which such civilizations come into existence times their average lifetime; just as the number of people in a country is equal to the birth rate in the country (people born each year) times the average age to which people live.[1] If, to pluck some numbers out of the air, one radio-communicating extraterrestrial civilization is born into the Galaxy every year, and such a civilization lasts a million years on average, there would be the potential for us to hear transmissions from a million radio-transmitting civilizations in our Galaxy.

To get a grip on the numbers, Drake assumed that the birth rate of extra-terrestrial civilizations in the Galaxy was related to the birth rate of stars: the more stars that are born, the more civilizations there will be. Implicitly, this assumes that the only places where intelligent, communicating civilizations exist are on planets orbiting around stars. Obviously, this is based on the example of our own Earth and Solar System, but so is most of the calculation. This limitation reflects the poor state of our knowledge about life elsewhere in the Universe.

The birth rate of stars in the Galaxy is the least controversial component of the Drake Equation, although this number still has uncertainties. The birth rate has changed over the life of the Galaxy, with more stars born when the Galaxy had more gas and fewer stars, early on in its 13-billion-year history. As its gas has run out and the stars that it made have energized the gas and made it more difficult for that gas to condense to form new stars, the birth rate has declined. One or two stars are born per year in the Galaxy now, but on average the Galaxy formed stars at the rate of about ten per year. Consistent with this figure are the number of stars in the Galaxy (about 100,000 million, most of them similar to the Sun, see chapter 10), and the lifetime of stars (about 10,000 million years for the Sun). This huge figure gives substance to the general feeling that the potential number of intelligent extraterrestrial civilizations living in our Galaxy could be very large.

Not all of these stars have planets, however. It could be that planetary systems are formed by some rare event. In 1917, the astronomer James Jeans (1877–1946) proposed such a theory. He hypothesized that, when one star passed close to another, it tore out a filament that condensed into planets. Because stars are a long way apart, near collisions are very infrequent, and, if what Jeans had conceived ever happens, it happens rarely. If this was the way that planets are made, the fraction of stars that have planets would be very small. Nowadays, astronomers think that planets are formed by a process that is intrinsic to nearly every star. Planets are a by-product of the formation of stars (chapter 2), and *most* stars will have planetary systems; at least, most stars of low mass like the Sun. Since solar-type stars constitute the majority of stars in the Galaxy, the fraction with planets is quite large; with the discovery of exoplanets, astronomers are now getting a good empirical idea of the fraction's actual size. Certainly, at least 10 to 20 per cent of solar-type stars have planets, perhaps as many as 40 per cent. On this reckoning, there are at least 10,000 million planetary systems in the Galaxy.

It does not seem likely that all of those planets could support life; they might be too cold, or have no solid surface to which life can anchor, or pose some other problem. What we do know is that life has evolved on Earth, so we can ask, 'How many Earth-like planets are there in each planetary system?' For our Solar System, the answer is that there are four terrestrial planets (Mercury, Venus, Earth and Mars). Wolszczan's star, the pulsar (see p. 18), has three Earth-like planets, although that star is atypical. On the other hand, many of the recently discovered planetary systems have a massive jupiter in the position in which Earth is found in our system, and it seems likely that if there were earths there they would have been destroyed in collisions with the much larger planet. The evidence points in both directions and – because their technology is not yet at a stage where Earth-like planets can be reliably detected – astronomers are still guessing at how many earths exist in typical planetary systems. To make some progress, let us estimate that a typical system has one Earth-like planet. This implies that there are 10,000 million Earth-like planets in the Galaxy.

The question 'On what fraction of earths does life in fact evolve, even in a simple form, such as bacteria?' is even harder to answer definitively. None of Wolszczan's planets is habitable, and this suggests the answer is zero, but we know the pulsar planetary system is not typical. In the example we know best, our own Solar System, life evolved on Earth and, as we will see, it might have evolved on Mars. Life probably has not developed on Mercury and Venus, because they are not 'habitable'. So this suggests that half or one-quarter of Earth-like planets has life. Is this more typical? Life began on Earth soon after it was born. This would imply that it is rather easy for it to start, in which case perhaps a more sizeable fraction of earths supports life. Dare we estimate that as many as 5,000 million earths – half of the 10,000 million in the Galaxy – have life on them?

Next is a critical factor in the Drake Equation: the fraction of earths on which life started that has gone on to give rise to intelligence. Again we have to use our planet as the typical case, since it is the only one we know. Defining what intelligence is in this context is not easy. If we mean 'life that lives co-operatively' we could be talking about some dinosaurs (which laid their eggs in crèches for mutual support or defence) or even bees. Many species on Earth have this facility, and thus intelligence might be common in the Universe. If we mean 'life that has started to use technology' we could be referring to birds (some of which manipulate sticks to get at food) or primates, such as chimpanzees (which not only use sticks, but also sharpen them to be more effective), or *Homo habilis* (one of the ancestors of *Homo sapiens*, who lived about 2 million years ago, and made the first stone tools). There are fewer, but still numerous, species on Earth with this facility. But this level of intelligence will not make a civilization that can communicate over interstellar distances. If we mean intelligence as advanced as the human brain, then there is only one example on Earth.

Fortunately, bringing the hard part of the question 'What is intelligence?' to an indefinite conclusion does not stop us taking a short cut to answer 'How often on earths does intelligence develop?' For much of the Earth's history there were only single-celled creatures, but in the last few hundred million years complex life has existed, and has created us. Capability and intelligence are such evolutionary advantages (or at least, they have been thus far) that we might suppose that life will always evolve to become complex and intelligent, given enough time.

On the Earth itself, the process that evolved complex life took nearly 4,000 million years. The Earth will last as a planet for at least another 4,000 million years. After that, the Sun will expand into a red giant and then die. The Earth may not survive this process at all; certainly it will not survive as an inhabitable planet, and it might become uninhabitable long before it is destroyed. As the Sun warms on its development into a red giant, it will roast the land and evaporate the oceans. The simplest calculations suggest that this will happen in 1,000 million years. This reckons without the ability of the Earth to adapt its environment to increasing warmth and to maintain its surface temperature (the Gaia hypothesis, see chapter 8). It also ignores the possibility that some other disaster might happen, earlier than the Sun expanding, to render the Earth uninhabitable (chapter 12).

To summarize, a planet similar to the Earth would be without intelligent life for the first 4,000 million years, and would then accommodate intelligent life for the next 1,000 to 4,000 million years. If we look at the entire collection of Earth-like planets in our Galaxy at arbitrary points during this typical history, the proportion that have intelligent life will be between one-fifth and one-half, but could be less if planetary disasters are commonplace. So perhaps there are 1,000 to 2,000 million planets in the Galaxy that sustain intelligent life. (But in chapter 16 I will give reasons to show that this might be a gross overestimate, and the real answer might be closer to one.)

Less uncertain is the fraction of intelligent civilizations that develops radio-telecommunications ability, if we assume that all intelligent life evolves to produce this technology in a relatively short period of time, as happened here on Earth. In which case, every one of those 1,000 to 2,000 million planets has a civilization that is potentially radio transmitting.

So there are 100 billion stars in the Galaxy and, on the above reasoning, 1 billion potentially radio-transmitting planets; that is, 1 per cent of stars could have a planet with a civilization that is transmitting messages to us. We already established that ten stars are born in the Galaxy per year, so the birth rate of potentially radio-transmitting civilizations is one every ten years.

In the above discussion, I have been very broad-brush in order to bring out the principles; authors who have gone into the various components of the equation in detail tend to end up with somewhat smaller numbers at this stage. But, as I will show, the inherent uncertainties of some of the components are very many indeed, and it does not seem worth trying to be more accurate about the less controversial ones. The Drake Equation is the best that we can do to be quantitative about the number of civilizations in the Galaxy with which we can communicate, and it has been very influential in focusing an agenda, not only for the meeting at which Drake originated it, but in the development of the new science of astrobiology. But the truth is that we are still far from using it to reach an agreed, accurate conclusion.

In fact, the final component of the Drake Equation is the most difficult one to estimate, simply because we have virtually no experience on which we can rely for an answer. For how long does an intelligent civilization transmit radio communications? The question was simplistically summarized as 'How long do they last?', the title of a scientific paper published in 2011 by Seth Shostak (b. 1943), the senior astronomer of the SETI Institute, a research organization based in California.

Some authors have suggested that we can look back into history and see how long, on average, a civilization lasts on Earth. Michael Shermer (b. 1954), a science writer and a historian of science, looked at the duration of sixty civilizations, including Mesopotamia, ancient Greece and Rome; the dynastic monarchies of China and Japan; a number of cultures from Africa, India, Central and South America; and selected modern European and American states. Altogether, the sixty civilizations endured a total of 25,234 years; therefore the average lifetime of a civilization is 420 years.

If this is the appropriate timescale to put into the Drake Equation, the number of extraterrestrial civilizations broadcasting to us in the Galaxy (the birth rate × the lifetime) is about forty, on my approximate figures, and could be a lot fewer.

At another extreme, one might argue that the appropriate lifetime for the Drake Equation is not the duration of individual political civilizations, but, since the knowledge of radio technology can be readily passed on from one civilization to another, the lifetime of technology over many civilizations. Once a species has developed radio technology, it might keep it for ever, even if the species groups into successive civilizations or evolves into higher forms. In that case, the lifetime of radio-transmitting planets could be a long time. The Earth is radio-transmitting and has about 4,000 million years left before the Sun swells up into a red giant. Maybe it will be habitable for only 1,000 million years of this time, in which case the appropriate lifetime for the Drake Equation is 1,000 million years. By this reckoning, the number of extraterrestrial civilizations broadcasting to us in the Galaxy would be about 100 million.

A more gloomy view could be that intelligent life brings about its own destruction pretty soon after it develops: degenerating and quickly losing technological and economic capability. This view was very prevalent during the era of the Cold War and under the threat of nuclear warfare, a danger that has not entirely passed. Concern might now be focused additionally on population growth and the prospect of mass starvation or catastrophic global warming. There again, even if the human race were to become extinct, intelligent life might reconstitute itself and repopulate the planet, perhaps learning from the experience of past species, retaining or rediscovering the knowledge of radio transmission and not making the same mistakes of self-destruction again.

An estimate of this key component of the Drake Equation, the lifetime of a radio-transmitting planet, is difficult, because we have only speculation, and our own prejudices about what might happen to humans over the millennia, to guide us. As the physicist Niels Bohr (1885–1962) is said to have commented, 'Prediction is difficult, especially about the future.' What the Drake Equation has told us is that the number of radio-transmitting civilizations in the Galaxy ranges from one (us) to 100 million.

The Drake Equation can be summarized in algebra as

$$N = R_* f_p n_e f_l f_i f_c L$$

Here N is the number of radio-transmitting civilizations in the Galaxy; R_* is the birth rate of stars in our Galaxy; a fraction f_p of which have planets; n_e earths per planetary system; f_l is the fraction of earths on which life evolves; f_i is the fraction that produce intelligent life; f_c the fraction of civilizations that develop radio-telecommunication ability; and L is the lifetime of an intelligent, radio-communicating civilization. An engraved brass plaque with this equation on it is screwed to the wall of the room at the National Radio Astronomy Observatory, Green Bank, West Virginia, where the meeting of 1961 was held, and where Frank Drake first wrote the equation on a blackboard.

In 1966, Soviet radio astronomer Iosif Shklovsky (1916–1985) and US planetary scientist Carl Sagan (1934–1996) used Drake's methodology to calculate the number of habitable planets. They 'guesstimated' the number of radio-transmitting civilizations in the Galaxy at 1 million. At a conference in 1971, however, Sagan noted that 'We are faced...with very difficult problems of extrapolating from, in some [factors], only one example and in [others] from no examples at all. When we make estimates we cannot pretend that these numbers are reliable.' Frank Drake himself estimated N at 10,000. But at a conference in London in 2011, astrobiologist Paul Davies (b. 1946) described the value of N as 'utterly moot', for the same reasons that have been sketched in my discussion above. It is an unsatisfactory position to be in after fifty years of hard work, and it is disappointing not to have any real proof of the answer that some people desire, that there are obviously many civilizations in the Galaxy with which we might be able to communicate. We simply do not know if that is so or not.

* * *

Are we alone? If it is difficult to calculate the answer to this question, what does the empirical Search for Extraterrestrial Intelligence tell us? The first search was initiated in 1960 by Frank Drake, using the 25-m radio telescope of the National Radio Observatory in Green Bank aimed at finding any extraterrestrial civilization that was beaming radio signals at us. The best radio frequencies for interstellar communication (the 'hailing frequency' as it is termed in *Star Trek*), Morrison and Cocconi argued in 1959, were between 1 and 10,000 MHz, where there is least interference from a planetary atmosphere and where radio noise from our Galaxy is also at a minimum. (Radio noise from the

Cosmic Microwave Background radiation[2] is also smallest in this band, but Morrison and Cocconi's paper was written before this potential source of interference had been identified.) The frequency that is transmitted by interstellar hydrogen, the most common element in the Universe, 1,420 MHz, is within this range. Drake's radio telescope was equipped with receivers that listened for transmission at 1,420 MHz, and it was convenient for him to listen at that frequency. Assuming that aliens in the intelligent, radio-transmitting civilization that Drake was aiming to find were like us, they would know about radio astronomy, and would realize that the people most likely to pick up their transmissions at large distances would have to possess large receiving dishes. The aliens would have built receivers to study interstellar hydrogen so they would guess that the people with whom they hoped to communicate would have receivers that covered 1,420 MHz. On that line of argument, one of reciprocity, it was reckoned that intelligent radio engineers on other planets would identify 1,420 MHz as the most likely frequency at which other radio engineers like Drake would be listening. So, following Morrison and Cocconi, Drake reckoned that 1,420 MHz was the most likely frequency at which he could find a radio communication from an extraterrestrial civilization. His search, based on this tentative syllogism, could justly be regarded as a long shot. But, Drake argued, the search has to start somewhere.

It was presumed that any signal would be transmitted in a narrow band, since this is the efficient way to pack information into the easiest signal to detect over large distances. Morrison and Cocconi went on to observe that any signal with narrow-band characteristics sent from an extraterrestrial orbiting planet to our own would drift in frequency, as a consequence of the Doppler shift. This is the change in pitch of a moving source of waves, just as the sound waves from a police car's horn or train's whistle change in pitch as the vehicle passes by. Because the speed at which the extraterrestrial planet is moving relative to ours changes all the time, so too would the frequency of the transmission. Drake set out to listen for signals with those characteristics from two nearby solar-type stars, Tau Ceti and Epsilon Eridani. He listened for 200 hours in April 1960 without success.[3]

Drake's pilot study, which he named Project Ozma – after Princess Ozma, the rightful ruler of the 'Land of Oz' in the books by L. Frank Baum (1856–1919) – was the model for more advanced programmes, with larger telescopes, more sensitive receivers and, most importantly, better signal processors, which took advantage of the great improvement in information technology since 1960.

Some of these subsequent searches were targeted, listening for long periods of time to promising stars. Others were wide-area searches, scanning strips across the sky at random. One such search, using the Big Ear radio tele-scope in Ohio State University Radio Observatory, simply looked up at a fixed angle above the southern horizon and let the sky drift through the sensitive beam of the radio telescope. Designed by radio astronomer John Kraus (1910–2004), the Big Ear was used from 1961 to study faint radio galaxies, but in 1972 the National Science Foundation (NSF) withdrew funding and the team was largely disbanded. Looking for a project that was not labour-intensive, Kraus, who had become interested in the possibility of finding radio-transmitting civilizations, started a SETI project. The equipment was automated and had to be visited infrequently – once every three days – mainly because the disc drive that held the data was full and needed to be read out to tape. In those days the state-of-the-art disc-drive held what was regarded at the time as a whopping 1 megabyte. Nowadays, most home computers have discs a thousand times bigger, and state-of-the-art discs can hold a million times more data. A crude summary of the data was printed on computer paper.

When scanning through the printout from 15 August 1977, computer scien-tist Jerry R. Ehman spotted the strongest narrow-band signal that he had ever seen (6). It occupied only one frequency channel of the fifty recorded by the receivers, so it counted as narrow band, and it lasted about a minute: exactly the time that it would take for a celestial signal to pass through the beam of the radio telescope. Its strength during the burst was coded as 6EQUJ5, namely 6, 14, 26, 30, 19 and 5 times the background noise; this was exactly the profile expected from a point-like celestial radio source passing through the telescope beam. Ehman scribbled 'Wow!' in red on the computer printout, and for ever afterwards the signal has been known as the 'Wow! signal'. It was never seen again nor identified with any known phenomenon: not a planet, asteroid or moon in the Solar System, not an artificial satellite or spacecraft, nor a ground-based transmitter. It is just possible that it was an illegally transmitted signal (the 1,420 MHz frequency is supposed to be a radio-quiet band preserved for astronomers) from a secret satellite or from a ground-based transmitter, the signal from which had reflected off a piece of space debris orbiting the Earth. Or it was a bug, a glitch, or some electronic hiccup. Or could it have been ETI, signalling once and never again?

SETI programmes continued, modestly funded by US government money through NASA and others, until Congress pulled the plug in 1993. Two private foundations, the SETI Institute and the SETI League, have continued the searches, as they have found it financially possible, using time rented from radio telescopes across the world, paid for by grants and donations. The finan-cing of the searches is a perennial problem, since grant authorities regard the projects as highly speculative, and prefer to fund science that is more likely to get a result.

Project SERENDIP (Search for Extraterrestrial Radio Emissions from Nearby Developed Intelligent Populations) is a University of California version of SETI that analyses radio signals from a receiver mounted in the focal plane of a radio telescope. It points to a location that has not been chosen by the SERENDIP project – the target is just off from the area that the radio telescope is focusing on for its main programme, whatever that is at the time. Its receiver in 2011 is known as SERENDIP-V. It looks over a wide-frequency band of 300 MHz, centred on 1,420 MHz, for signals transmitted by extraterrestrial intelligence; perhaps aliens do not use the 1,420 MHz frequency, as Morrison and Cocconi suggested they would if they were like us. The receiver is mounted on the Arecibo radio telescope in Puerto Rico, currently the largest in the world. The Arecibo telescope is aging and its funding is under pressure; if it closes, the equipment can be taken elsewhere, but SETI will suffer a setback.

Another, similar, project uses a receiver that points to seven places in the focal plane of one telescope, and analyses the data according to Morrison and Cocconi's suggestion that aliens would transmit in a narrow band, within a few MHz of the frequency band of hydrogen. This project, also based at the University of California at Berkeley, is called SETI@home. The data is analysed using the spare capacity of home computers, which adds up to considerable computing power. This all means that the project is very sensitive to faint signals – effectively, the project searches a big volume of space, with a correspondingly greater chance of success. SETI@home was one of the first 'citizen science' projects, like the Kepler space project (p. 21), amassing aid from thousands of volunteers ('crowd sourcing') in order to process a large amount of data. One ETI signal, a source called SHGb02+14a, caused a brief flurry of exciting messages on the Internet in 2004. Its narrow-band signal had a frequency drift, as predicted by Morrison and Cocconi, but at a rate that corresponded to a planet moving forty times the speed of the Earth round the Sun. There are no nearby stars in the direction of SHGb02+14a. SETI@home scientists do not support the conclusion that the signal was significant, and put it down to noise, like Tesla's 'dot, dot, dot'.

* * *

Of course we need not sit shyly and silently, like wallflowers in a galactic ballroom, and wait to be approached. If we want to communicate with extraterrestrial intelligence we can send out signals ourselves, for example by following the protocols articulated by Morrison and Cocconi. There have been a number of attempts to contact intelligences in other planetary systems, a process known as 'Active SETI'.

The first such messages were letters rather than broadcasts. Anodized aluminium plaques were attached in 1972–73 to the Pioneer 10 and Pioneer 11 spacecraft, which were planned to travel out of the Solar System. Each plaque showed a line diagram describing its origin, in case it was found by extraterrestrial space travellers. The Voyager 1 and 2 spacecraft, launched in 1977, contained a Golden Record – a phonograph record made of gold-plated copper in a cover – containing sound messages and photographs, as well as a diagram much like the Pioneer versions. All these spacecraft are now well beyond Pluto. The Pioneer plaques (7), designed by planetologist Carl Sagan and radio astronomer Frank Drake, with the artistic participation of Linda Sagan (b. 1940), show:

- A diagram of the change in the hydrogen atom that creates the 1,420 MHz frequency, 21-cm wavelength radio waves from hydrogen clouds in interstellar space. This is intended to show the scale of some of the other components of the picture.
- Figures of a man and a woman. Both are naked and they are intended to be of no particular race. The man holds up his hand in greeting. Their height is shown in binary notation in units of 21 cm.
- A star-like pattern of radial lines of different lengths in different directions that shows the relative position of the Sun, the centre of the Galaxy and fourteen pulsars. The pulsars are identified by their periods (enumerated in binary arithmetic, in terms of the period of the hydrogen radio waves). Provided the extraterrestrials know the pulsars in the Galaxy, this shows where the spacecraft came from, and, since the periods of pulsars change, when it was launched.
- A diagram of the Solar System, with the Sun and the nine then-identified planets from Mercury to Pluto, showing the spacecraft leaving Earth and exiting the Solar System via Jupiter.
- A simple silhouette of the spacecraft is shown behind the humans, again intended to show their scale.

Of course we do not know when, or whether, these spacecraft will be found. We do not know if their finders will have, or be able to make, the right technology to play the record, or the empathy to understand the diagrams and messages. Neither do we know if extraterrestrial intelligence will hear or understand our radio broadcasts to them. In 1974, the Arecibo radio telescope broadcast a message to a globular cluster of stars in the constellation Hercules, known as Messier 13, at a ceremony to mark an upgrade of the telescope.[4] The message was a fun way to demonstrate the capability of the newly installed equipment. It was a picture of 1,679 pixels, a number chosen because it is semi-prime, the product of two prime numbers: 73 × 23. Because of this, a picture can be formed in just two ways – 73 rows by 23 columns, or 73 columns by 23 rows – but only the first makes a sensible image (8). This picture was also created by Drake and Sagan. The seven icons in it depict the following:

- The numbers 1 to 10 in binary format, to establish the notation system used.
- The numbers 1, 6, 7, 8 and 15 in binary, which are the atomic numbers of the elements hydrogen, carbon, nitrogen, oxygen and phosphorus. This is intended to convey that we are carbon-based life forms, with a biochemistry based on these elements: the ones in our DNA.
- The chemical formulae for the sugars and bases in the nucleotides of DNA, showing the fundamental basis of our biochemistry.
- The number of nucleotides in DNA, expressed in binary, and a diagram of a double helix.
- A diagram of a person, with the typical height of a human expressed in binary as a multiple of the wavelength of the radio transmission, and the population of the Earth.
- The person 'stands' on the Earth in a simple diagram of the Solar System, with the nine planets, from Mercury to Pluto (considered a planet at that time), intended to convey where we are.
- A diagram of the Arecibo radio telescope and its diameter, again in terms of the wavelength of the transmission, to indicate what we are doing.

Messier 13 is 24,000 light years away, and was chosen because it was in the sky during the ceremony. We will not know for nearly 50,000 years if anyone on a planet there has received the message, decoded it and replied. A similar ceremonial transmission in 2008 consisted of the song *Across the Universe* by the Beatles, beamed by a tracking station of NASA's Deep Space Network (DSN) towards the star Polaris to mark the forty-fifth anniversary of the DSN, and the fiftieth anniversary of NASA. Polaris is 430 light years away and is a bright star of advanced age. If it has a planetary system, it is probably uninhabited, so few astronomers expect a reply.

The Interstellar Rosetta Stone is a more comprehensive and longer scientific message, created by Canadian scientists Yvan Dutil (b. 1970) and Stéphane Dumas, and built mainly along the same lines as the Arecibo message. It was transmitted in 2003, under the name 'Cosmic Call', by a large Russian radio telescope to five nearby stars, after a smaller pilot experiment in 1999. It contains individual messages of varying profundity from thousands of people in more than fifty countries. It includes representations of the flags of the world, and David Bowie's song *Starman*. It also contains the text of a resolution passed by the New Mexico state legislature in 2003, designating the second Tuesday in February as 'Extraterrestrial Culture Day' in that state (it is celebrated primarily in Roswell, New Mexico) and including the intention, now fulfilled at least in part, that the resolution should be 'transmitted into space with the intent that it be received as a token of peace and friendship'. A few other attempts have been made to send interstellar messages, but these were more in the nature of light-hearted stunts than thought-through attempts to communicate with extraterrestrial intelligence.

These trial attempts at Active SETI provoked a debate about how wise it is to advertise our existence and that of the nice planet we inhabit. The message might amount to an invitation to come and stay. A look at the history of our own species within Earth – from our encounters with, competition with, and eventual extinction of the Neanderthals about 25,000 years ago, to the displacement of native Americans by European pioneers, to quote only two examples – should cause us to reflect on what might happen should our invitation be taken up. We have, however, already given away our location unintentionally, by the signals radiated into space by radio and television transmissions, so it is probable that no extra harm was done.

* * *

No SETI project has definitely detected that elusive first extraterrestrial signal yet, whether sent to us deliberately in reply to our messages or unprompted.[5] Astrophysicist Paul Davies (b. 1946) calls this the 'eerie silence'. What will be the effect, if and when the eerie silence is broken? What will it mean, if and when we discover that there are intelligent people in the Universe other than us? This is a human and philosophical question that has been considered for millennia, since the days of ancient Greek philosophy. Historical discussions still have relevance for us today; scientific and technical advances have not changed underlying questions about the nature of ourselves as human beings. Are we special or are we typical? Are we a part of the Universe or are we apart from the Universe? Are we unique?

Greek philosophy is organized common sense, the systematic development of attitudes of mind, the beginning of science. But, through systematic thought alone, Greek philosophers failed to decide on the number of worlds in the Universe and whether any are inhabited. They came to completely opposite conclusions. According to Epicurus (fourth century BC):

There are infinite worlds both like and unlike this world of ours. For the atoms being infinite in number, as was already proved, are borne on far out into space. For those atoms which are of such a nature that a world could be created by them or made by them, have not been used up in one world or in a limited number of worlds.... So that there nowhere exists an obstacle to the infinite number of worlds.

But according to Aristotle (also fourth century BC):

Either, therefore, the initial assumptions must be rejected or there must be only one centre and one circumference; and given this latter fact, it follows from the same evidence and by the same compulsion, that the world must be unique. There cannot be several worlds.

Epicurus, like his predecessors – the philosopher Leucippus and his pupil Democritus (fifth century BC) – and the Roman poet Lucretius (first century BC), who popularized Epicurus's work, thought that small, simple, indestructible particles are the basic components of the entire Universe. This theory is known as 'atomism', and the particles are atoms (from the Greek, meaning 'indivisible'). The atomists were concerned to understand the nature of change in the world, and to determine whether it was real or an illusion. They believed that everything we see is made of atoms that can move around and arrange themselves in different ways in space. According to the atomists, atoms are disposed in space, the 'void', which is infinite. The different possible arrangements make up the various things that we can feel, see, hear, smell, taste and, in general, sense.

In his analysis of the problem as to whether there was one world or many, Epicurus started from the idea of the infinite size of space and therefore of the number of atoms in it. There were many ways that the atoms could be arranged, and some of them would repeat. Some arrangements would be completely different from any we know, others would be somewhat unfamiliar and some would be similar to the arrangements of atoms in our world. Epicurus thus came to the conclusion that our world is one of many.

The philosophy of the atomists came under considerable scrutiny and criticism. The Greek philosopher Plato (fifth–fourth century BC) was repelled by its randomness, and suggested that atoms merely jumbling up with other atoms could never produce the beauty and symmetry we find in the world. Perhaps there were basic corpuscles (minute particles) like atoms, but they had fundamental geometric shapes. Corpuscles of these shapes, Plato said, made up four elements: earth, air, fire and water.

Plato's student Aristotle took up his teacher's concept of the four elements, and rejected atomism, or anything like it. He suggested instead that the world was indeed made of the four elements – earth, air, fire and water – but that they were not atomic and could be indefinitely subdivided into finer and finer pieces. He suggested that change occurred when a material substance realized its potential, much as a piece of clay could become a pottery cup. The potential of a substance influenced the way that it changed. A stone, for example, had the potential to be a part of the world and would realize this potential by falling. It was its nature to fall.

Aristotle postulated that there was one more element, the quintessence (meaning 'fifth element') also called ether, a divine substance that makes up celestial bodies, such as the stars and planets. (The 'planets' at this time meant the Moon and the Sun, as well as Mercury, Venus, Mars, Jupiter and Saturn. These are the seven moving or 'wandering' celestial bodies, distinct from 'fixed' stars, which stay the same, relative to each other.) Ether was fundamentally different from the other elements. According to Aristotle, the four terrestrial elements tend to move towards their natural place – air floating in the sky and earth falling downwards towards the centre of the Universe, for example – but by contrast, the ether remains in the heavens and never falls. Thus our world, the Earth, is at the centre of a series of concentric crystal spheres carrying the Moon and the Sun, five planets (as we would call them now, namely Mercury, Venus, Mars, Jupiter and Saturn) and the stars. The celestial bodies move eternally around the Earth with unchanging circular motion. Aristotle was inevitably led to the idea that there was just one centre to the rotating Universe, and that therefore our world was unique.

Greek civilization gave way to Roman rule, and the western Roman Empire itself fell in the second half of the fifth century. Europe became Christianized, and astronomical theories were adapted to Christian thought. The link between Aristotle's beliefs and the writings of the Bible was forged in the thirteenth century, by intellectuals at the University of Paris. They were known as Thomists, because they were led by a Dominican friar, Thomas Aquinas (1224/25–1274), who was later canonized as St Thomas. The Thomists believed in a 'chain of being' that stretched from our changeable world up to the eternal celestial bodies, mounted in crystal spheres. Moving outwards from our world, the planets progressively became more perfect. The closest planet, the Moon, was a perfect sphere but it changed its shape and had grey patches (the features known in folklore as the Man in the Moon). The Sun was thought to be spotless. The planets, from Mercury, Venus, Mars and Jupiter to Saturn, moved increasingly slowly, indicating a progressive approach to eternal lack of change. Beyond the celestial spheres, it was thought, lay the completely motionless, utterly perfect, unchanging, eternal Empyrean, the dwelling place of God, which the Thomists identified with heaven.

In this picture, God encompassed our world, surrounding it and looking after us, much as a father would look after his children. His love for us, the inhabitants of the world, was so great that He came to the world in the unique incarnate form of His Son, Jesus, whom He allowed to be put to death in a particularly cruel manner in order to save us from our sins and make possible our transition from the mutable world of corruption and decay to eternal life in heaven. We human beings were thus seen as the particular focus of God's attention. This philosophy became the accepted teaching of the established Church, with the obvious conclusion that there was only one world, at the centre of the Universe (9).

There was a particular Christian theological argument that backed up the astronomical argument that other worlds and extraterrestrial intelligence do not exist. If there were many worlds, inhabited by intelligent beings, then for these beings to be 'saved' and to have the opportunity to enter heaven, it seems fair that they should have been offered the incarnation of the Son of God to bring them God's message. This raises the question of how often Jesus, or his equivalent, should have been born and executed in a despicable manner on world after world. The proposition that the Son of God should have suffered innumerable times in all the other possible worlds in the Universe was repugnant, and was another argument for the unique status of the Earth and human beings.

Aristotle's world picture, elaborated by the Thomists, held sway for centuries, but fell with the scientific investigation of the orbits of the planets around the Earth (as it was thought). These orbits were not perfect circles after all, but complicated hierarchical orbits of circles, the centres of which themselves moved in circles; such a system was called 'epicyclic'. The epicyclic system that was developed to describe the motions of the planets proved to be inadequate. Searching for a more accurate system by which to calculate the planetary orbits, in 1543 the Polish cleric Nicolaus Copernicus (1473–1543) published his theory that the Sun was the centre of the Solar System. The planets, including our world, orbited the Sun. This brought out the possibility that our world was not unique, since it was a planet like others. Copernicus himself did not go so far as to make this explicit link, but his followers did.

The Italian monk Giordano Bruno (1548–1600), who had once been a Dominican, shrugged off his order so that he could propose more radical ideas than it tolerated, and took up the logic of the Copernican theory. In 1584 he came to England, and arrived in Oxford to put forward his idea that there were numerous worlds. The Sun, he suggested, was a star. True, it was much brighter than the other stars, but that was because the Sun was far closer to us. There could be innumerable stars that stretched off into the distant Universe, each with planets like ours. Bruno wrote down his ideas in a book, *De l'infinito universo e mondi* ('On the infinite Universe and its worlds'), published in 1591. The then Archbishop of Canterbury derided him for supporting 'the opinion of Copernicus that the Earth did go round, and the heavens did stand still; whereas in truth it was his own head which rather did run round, and his brains did not stand still.' Bruno's views on this cosmology and on religion were his downfall. Lured back to Italy by a nobleman who requested to be taught by him, Bruno was arrested by the Inquisition in 1592, and never saw freedom again. He was tried for his heretical theology, and burnt at the stake in Rome in 1600, 'his tongue imprisoned because of his wicked words'. This contemporary description of the way that Bruno met his death is not explicit, and we can only guess at what the words mean – imagining what they hide only makes the horror of the event worse.

Bruno's works were published at the time that the astronomer Johannes Kepler (1571–1630) was a student at the University of Tübingen. It is not known whether Kepler studied Bruno's works directly, but undoubtedly Bruno's ideas would have been discussed by Kepler's fellow students. Kepler gradually came to the conclusion that the Moon, the Sun, the planets and even the stars were worlds like ours, and might be inhabited. He argued that the dark spots on the Moon and the uneven nature of the boundary between its bright and dark sides implied that its surface was rough and mountainous like the Earth. He went on to speculate that 'there are on the Moon living creatures, with by far a larger body and hardness of temperament than ours to be sure, because if there are any there, their day is fifteen of our days long and they endure both the indescribable heat and the vertical rays of the Sun'.

Kepler's work took him to Graz and Prague, where he continued to study astronomy, particularly the motions of the planets. He concluded that the planets moved in ellipses, not circles, nor circles upon circles. Without circular motion there could be no unique centre of the Solar System, or of the Universe. This destroyed Aristotle's link between the uniqueness of the Earth and its position, and further weakened the claim that ours was the only world.

The Italian scientist Galileo Galilei (1564–1642), in 1610, showed just how credible Bruno's idea of multiple worlds was, by using the telescope to discover that there were moons in orbit around other planets. Jupiter, he found, had four large moons that revolve around it, just as the planets revolve around the Sun. He used the superior light-gathering of his telescope over the power of the naked eye to see that there were stars beyond those known since ancient times, just as Copernicus and Kepler had speculated. Galileo viewed the Moon with the telescope and confirmed Kepler's conclusion that it resembled our world in form; it had valleys and mountains, and what appeared to be seas (although we now know the dark patches on the face of the Moon are desert plains of grey lava covered with dust).

At this time in the history of astronomy, 400 years ago, it became established that there were in fact other worlds like our own. This discovery moved the possibility of the existence of alien life onto firmer ground, literally and metaphorically.

Kepler incorporated Galileo's new discoveries into his ideas about life on other worlds. He had been working on a book about the Moon since he was a student, which was published posthumously, in 1634, with the title *Somnium (Dream)*. It is a work of scientific imagination; it has been said that it was the first book of science fiction. Kepler wrote of an imaginary voyage to the Moon, describing its surface and the creatures that might live on it. Because a night and a day on the Moon are equivalent to one month on Earth, he reasoned that the lunar daily cycle is one of extreme heat and extreme cold, mitigated a little on the side that faces Earth, because the Earth warms the Moon. Kepler knew that Galileo had seen with his telescope that the surface of the Moon has a terrain like the Earth's, but with its scale exaggerated because of the low gravity: very high mountains, very deep fissures, valleys and craters. Likewise, Kepler imagined the lunar inhabitants to grow monstrous in size. He described them as nomads, roving over the surface of the Moon to seek refuge from the heat of the lunar days, some using legs, some wings and some floating in boats on the seas, looking to cool down in deep, shady caves.

Kepler's book is a precursor to this one, and any book on the modern subject of the science of 'astrobiology': the study of the distribution of life in the Universe (or the study of things that do not exist, according to some sceptics). With the development of space exploration by satellites and spacecraft, it is a science that has become timely, one into which both astronomers and biologists are putting a lot of effort. The bulk of the science is new, but its roots go back 400 years to when Kepler used scientific reasoning to imagine the biological conditions on the Moon and the consequences for anything that lived there. He imagined how the properties of the Moon – its gravity, the lunar climate and the terrain – would produce life to suit, a life that was different from that on Earth. He described, as I am going to attempt to do, the interrelationship between life and its cosmic environment.[6]

To answer the question, 'Is there life out there, and if so, where can we expect to find it?' we must begin with an understanding of life in the environment of our own Solar System, including our host planet.

4

Life in the Solar System

What we know for certain about the cosmic environment of life is that we ourselves, and the rest of the life-forms on Earth, live on or near the surface of the planet. The key word here is 'surface'. When we say that life lives on Earth, it implies that it occupies the whole of the planet. Scientists enshrine this concept in the use of the term 'biosphere', which means the entirety of the part of the planet that contains life. The word 'biosphere' suggests that life exists throughout the spherical volume of the Earth. But, mathematically, a sphere is also a surface, and the biosphere is a thin layer enclosing the surface of the Earth – life does not survive in the Earth's core, nor in the atmosphere far above the ground. When mathematicians talk about the volume *of* a sphere they actually mean the volume *inside* a sphere. In the mathematical ideal, a sphere is an infinitely thin surface. The biosphere, it is true, is not infinitely thin, although in proportion to the size of the Earth it is like the stretched rubber skin of a beach ball. Life on Earth resides in what Edinburgh-based astrobiologist Charles Cockell (b. 1967) has suggested could be called the 'biofilm', a neologism that stresses how thin the biosphere is. The Earth's biosphere is just a thin skin that lies between the lithosphere (the sphere of rocks and ores that constitutes the solid Earth) and the atmosphere (the sphere of gas that surrounds the lithosphere).

Living within the biosphere as we do, and with the perspective of a small creature under a vast blue sky, the biosphere seems to us to be large and enveloping. From space, astronauts perceive the biosphere's true size in relation to the entire planet. ESA astronaut Ulf Merbold (b. 1941) described his experience when flying to the Mir Space Station in 1994: 'For the first time in my life I saw the horizon as a curved line. It was accentuated by a thin seam of dark blue light – our atmosphere. Obviously this was not the ocean of air I had been told it was so many times in my life. I was terrified by its fragile appearance.' NASA astronaut Loren Acton (b. 1936), flying in the Space Shuttle in 1985, was also struck by its fragility: 'Looking outward to the blackness of space, sprinkled with the glory of a universe of lights, I saw majesty – but no welcome. Below was a welcoming planet. There, contained in the thin, moving, incredibly fragile shell of the biosphere is everything that is dear to you, all the human drama and comedy. That's where life is; that's where all the good stuff is.'

If we pick apart the biosphere of Earth, could it help us to find a similar environment elsewhere in the Universe? Depending on exactly how it is made, life here needs some (but not all at the same time) of the following: a solid surface on which to anchor; bodies of water (lakes, rivers or seas) to provide a fluid medium in which to live or to replenish an internal supply of liquid that will carry our biochemicals into proximity with each other, so that they can react and sustain life; sources of energy (usually sunlight or tectonic heat) that can be used to drive the biochemical reactions that make life work; and an atmosphere of gases that are both absorbed and produced by living bodies.

These life-supporting characteristics are phenomena of the surface of a rocky planet, or somewhere near to the surface, like an ocean or the lower atmosphere. The exact properties depend on where the planet is in its solar system, and on its intrinsic nature, which in turn also depends on its location as well as its origins and history. In this chapter, I consider what the distribution of life in a planetary system might be, by using the only example that we know well: our Solar System.

Living organisms are composed of chemicals based on the element carbon. This is such a common property that chemical compounds based on carbon are called 'organic' chemicals, whether or not they are made by living creatures. The carbon atom has four positions at which it can bond with other atoms – which is more than most atoms – and the bonds can be of several types and strengths. This means that carbon can join together in chains and rings that are able to build up into very complex molecules. Complex molecules have the potential to interact chemically in a multitude of ways. The chemistry of life, biochemistry, is extremely complicated. After all, biochemistry is ultimately responsible for Leonardo da Vinci's painting of the *Mona Lisa* and Beethoven's Ninth Symphony, not to mention Einstein's theory of relativity. Carbon atoms have the potential to do the complicated things that constitute life.

Is carbon the only element that can form the basis for life? There are about a hundred elements known to chemistry. An atom of one element differs from the atoms of others by having a different number of protons in its nucleus. If the atom is isolated and undisturbed (by being over-energized – in a hot gas, for example – or by being part of a chemical compound) it will have the same number of electrons in orbit around the nucleus as the nucleus has protons. The element hydrogen has one electron and one proton; helium, two of each; lithium, three; beryllium, four; boron, five; and carbon, six. Beyond carbon are nitrogen (seven); oxygen (eight); and elements with more numerous electrons and protons, such as sulphur (sixteen), iron (twenty-six), and so on, up to hahnium (one hundred and five). There are no gaps, and chemists have a good knowledge of the properties of every element up to hahnium. Those beyond that (and some of the elements below it) are radioactive, and do not last for any length of time, so they cannot play a part in sustaining life.

Thus it is almost certain that we know of all of the chemical elements that exist. None of them has the same complex properties as carbon (although some, such as silicon, imitate carbon to a lesser extent). It is highly likely that carbon is the only element that can build complex molecules of the sort that biochemistry needs. The complexity of organic molecules is the reason why life takes complicated and varied forms; it is the basic chemical explanation for why carbon-based life is the most interesting kind – or at least, that is a good working assumption.

*   *   *

Carbon chemicals cannot carry out any biological function unless they have a medium to work in: water. Water is made of hydrogen and oxygen. The water molecule consists of three atoms in a line. The one formula that everybody remembers from school chemistry is the formula for water, H_2O, each molecule being a central oxygen atom with a hydrogen atom locked on to each side: H-O-H. The molecule is not a straight line; it is crooked, bent at an angle of $104°$ at the central oxygen atom. Hydrogen and oxygen are two light elements that are abundant in the Universe, so water is a common molecule, in cosmic terms. But it is a molecule that has uncommon properties. In spite of its lightness, water readily remains a liquid. This is because the electrons in the molecule are not evenly distributed. The oxygen atom has slightly more than its fair share of electrons, so it has negative charge. The hydrogen atoms have correspondingly less than their share of electrons, so they are positively charged. This means that water molecules like to nuzzle up to each other, with the negative charge in one molecule attracted to the positive charge in another. Four water molecules nuzzle up to each other in a tetrahedral shape. The force with which the molecules are attracted is sufficient to maintain water as a liquid over a wide range of temperature, but weak enough to allow other molecules to interpose themselves. This makes water a good solvent for many chemicals. There is no other common molecule with water's combination of properties that make it so suitable as the medium for biochemicals to get together and operate as life.

The existence of water on the surface of our planet is primarily the result of Earth's distance from the Sun. Sunlight, and the infrared radiation that is associated with it, warms the surface of our planet. If we were considerably closer to the Sun, our planet would be too warm; water would turn to steam and it would be unable to act as a solvent for biochemical reactions. On the other hand, if we were further away, water would be ice and unable to carry chemicals, because ice, being frozen solid, is nearly static. There is a zone in our Solar System within which water could readily exist on the surface of any planet located there. It is called the 'habitable zone' or, more informally, the 'Goldilocks zone', named after the fairy-tale. Like one of the bowls of porridge that Goldilocks discovered on the breakfast table of the three bears, the temperature in the habitable zone is not too hot, nor too cold, but just right.

The concept of the habitable zone works only because the orbits of the planets in our Solar System are approximately circular: the entire orbit of a planet will lie either fully inside or fully outside the zone. Some of the planets that orbit other stars have much more eccentric orbits. These planets might roast when nearest to their sun and freeze when furthest away. During the 'year' in which such a planet orbits its sun there would be extremes of climate, a very hot summer and a very cold winter. It seems likely that, although such a planet would, in principle, be 'habitable' twice in its 'year' between these extremes, it would not, in practice, be inhabited.

Calculating whether a planet is in the habitable zone for a given star depends not only on the distance of the planet from its star but also on the kind of star; how much of the stellar radiation is reflected back from the planet out into space (which in turn is dependent on cloud, water, ice and vegetation cover); whether the planet has an atmosphere; and the scale of its greenhouse effect.

The greenhouse effect holds energy inside an atmosphere, which acts as a sort of blanket. In our own situation, solar radiation is mostly sunlight, because the surface of our Sun is so hot: its temperature is some 5,800°C. Earth's atmosphere is, on the whole, transparent to sunlight. Our Sun therefore warms the ground of our planet and the surface of the sea, the temperatures of which rise to perhaps as much as 40°C, and the surfaces radiate infrared radiation, to which the atmosphere is, by contrast to sunlight, not very transparent at all. Water vapour and carbon dioxide in the air absorb infrared radiation, so the air warms up, providing an insulating blanket that traps heat inside.

The ability of an atmosphere to provide a greenhouse effect depends on its composition; the more carbon dioxide (for example), the more intense the greenhouse effect, the more incoming radiation that is trapped, and the hotter the surface of the planet becomes. Carbon dioxide is not the only greenhouse gas, not all carbon dioxide is man-made, and there are a number of other factors that have an effect on climate (such as the behaviour of the Sun), but the greenhouse effect is the scientific basis for discussing whether an increasing level of man-made carbon dioxide in our planet's atmosphere is responsible for global warming. This is a politically controversial issue with some complicated scientific aspects. Likewise, the calculation of the depth of the habitable zone in a planetary system is also somewhat complicated. In our Solar System, the habitable zone extends, perhaps, from the orbit of Venus to the orbit of the Main Belt of asteroids, located between Mars and Jupiter.

The concept of the habitable zone of a planetary system not only assumes that extraterrestrial life has the same requirements as terrestrial life, but also ignores the possibility that a planet's individual properties might produce the right (or wrong) conditions for liquid water to survive. A planet that lies far from its sun might have such significant sources of heat as volcanic activity (which directly releases heat), lunar tides (which work the oceans and, indeed, the malleable volumes of rock of the 'solid' mass of the planet, causing them to warm up) or radioactive decay (which directly releases heat, but also energetic particles that yield heat when they are absorbed by surrounding material). If such a planet has water, it may be liquid even if the planet is outside the habitable zone, in an orbit where one would expect its water to be solid ice. Water may be liquid beneath a frozen icy surface, warmed from below by a planet's hot core. A planet featuring such conditions, although located outside the habitable zone, could have a limited environment that supports life. This is called a 'habitable niche'.

The concept of the habitable zone also ignores the fact that the temperature of a planet varies across its surface. The temperature of our own Earth is, typically, below the freezing point of water at its poles and at the summits of its high mountains. And although the temperature in the hottest regions, near the equator, is never above the boiling point of water, liquid water quickly evaporates from the deserts of Saharan Africa, Central Australia, Chile and the American Southwest, to name only a few places. On small planetary bodies or moons there is no atmosphere, because their force of gravity is so weak; in such circumstances ice evaporates quickly without ever melting into water first. The technical expression is that ice 'sublimes' directly into vapour. Therefore, even if a small body were in the habitable zone and had water, that water might dissipate quickly, and the planet would become dry and lifeless.

* * *

In our Solar System there are, at the present day, reckoned to be eight planets, three hundred planetary satellites, millions of asteroids and very many comets. Bearing in mind that the concept of the habitable zone of our Solar System is rather ill-defined, we can nevertheless say that two planets (Earth and Mars), three moons (ours and two Martian moons) and some asteroids on more or less circular orbits lie in the habitable zone, and we know that life exists on at least one of these bodies. Is it conceivable that there is water and thus life on any of the rest? The short answer is yes, and in fact water has been detected or inferred in several places, not only in the habitable zone of the Solar System (beyond Earth) but also outside it, in some habitable niches.

The inner Solar System consists of four planets that orbit closest to the Sun and are called the terrestrial planets: Mercury, Venus, Earth and Mars. They are all similar to Earth: each is small, with a rocky surface and an atmosphere – a very thin atmosphere in the case of Mercury, because it is both small (and its gravity is weak) and near the Sun (and therefore hot), so that molecules of its atmosphere readily escape. Well outside the habitable zone, scant atmosphere, and a surface so hot that tin melts: Mercury is not a promising place for us to find life.

Venus is next out from the Sun. The planet is Earth-sized and has been able to retain a dense atmosphere of carbon dioxide, under which is a hot, dry, volcanic surface. The surface has not been investigated directly, except a few small areas that were pictured from some Russian landers. NASA's Magellan spacecraft has, however, made a radar survey of the terrain of Venus through the opaque cloud cover, at such detail that we can see conical volcanoes, lava flows and ash plains. There is no sign of water – either now or recently – no lakes, seas, rivers or dry river valleys. If Venus did once have water, the heat of the Sun on the planet's surface would certainly have vaporized it. This process made the surface temperature rise further, via the greenhouse effect, which weakened the crust and made Venus extremely volcanic, erasing any signs left by the water that the planet once had. Out poured carbon dioxide, which augmented the greenhouse effect, and so on. This runaway effect makes the surface temperature of Venus about 460°C; any water has long since evaporated entirely from the surface of the planet. Venus is no more promising as an abode of life than Mercury.

The next planet outwards is the Earth. The life here is our only model for finding signs elsewhere, and our planet thus deserves the detailed consideration that it will get later on in this book.

Mars is promising. In our Solar System, it is the planet that is the closest to and most like Earth. It is inside the habitable zone – on some calculations – and, since the surveys of Mars by the Viking Orbiter spacecraft in the late 1970s, it has been known that at some time in the past Mars had abundant liquid water on its surface. There are some indications that there may be water on Mars still. It seems quite possible that life could previously have developed there, and may yet survive in some limited ecological niches. The evidence is quite extensive, and Mars as a potential abode for life also deserves a chapter to itself (chapter 13).

Beyond the four inner planets are four more – well outside the habitable zone – that are much larger, with thick gaseous atmospheres, deep within which there may be a solid core. They are called the gas giant planets: Jupiter, Saturn, Uranus and Neptune. These planets have no solid surfaces as such, and are very cold. They have lots of water, but it exists as ice or vapour. If there is life on these planets, it is very alien indeed.

* * *

It is not a coincidence that the boundary between the terrestrial planets and the gas giants corresponds to the outer edge of the habitable zone of the Solar System. Within this outer edge, ice in the nebula that formed the planets melted and evaporated, releasing gases trapped within. What was left was so-called refractory material: solid bits of dust, which accumulated, in the early history of the Solar System, into the terrestrial planets. Astronomers call this outer edge the 'snowline' – an analogy with the contour on a mountain above which it is always snowy – because beyond this line gases in the nebula that created the planets remained solid, as ices. Here, all materials, including ices, gases and solids, accumulated into the larger gas giant planets. Inside the snowline, however, only the solid, dry material built up into planets. But if the heat of the Sun dried everything out, why is it that water can now be found in abundance here on Earth, and possibly elsewhere in the habitable zone?

The answer is that water was brought to the inner regions via the Solar System's transport-and-delivery system. There are numerous small, water-rich bodies that orbit between the planets. These are called asteroids, comets and Kuiper Belt objects, and include what was formerly called the ninth planet, Pluto. Many have orbits that cross the orbits of the larger planets, and there is an ever-present danger of collision. Asteroids or comets that collide with a planet are absorbed. This is how large quantities of water were delivered to the Earth and to other planets.

The impact scars of asteroid and comet collisions show clearly on the faces of the planets and moons that have solid surfaces. They are the craters that completely cover the surface of the planet Mercury, the Moon and other rocky satellites, but they are found in lesser numbers on all the rocky planets, including Earth.

The larger bodies of the Solar System have absorbed material from many comets and asteroids. Those smaller bodies that remain are now orbiting in three relatively safe zones of the Solar System. Some asteroids orbit in what is called the Main Asteroid Belt, which lies between Mars and Jupiter. Comets orbit slowly, in a hypothetical spherical zone beyond the far reaches of the Solar System known as the Oort Cloud. From time to time some comets interact with one another, or maybe with a passing star or interstellar gas cloud, and are disturbed from their slow, cold, comatose existence, falling in towards the Sun, passing through the planets. An intermediate zone of the Solar System – a disc extending from the outermost planet, Neptune, to the Oort Cloud – is known as the Kuiper Belt, in which orbit small, cold bodies, a mixture of asteroids and comets.

What can be said about the possibility of life on these bodies? It is not likely that it exists on any of them, but nevertheless they do play a key part in the story of the development of life elsewhere. Not only do they deliver water to planets and satellites, but they also bring the seeds of life, in the form of prebiotic chemicals. These biological chemicals, assembled in the right way, make life. Here's how.

* * *

Comets are icy bodies left over from the formation of the Solar System, which found themselves kicked out great distances from the Sun after interacting with the gravitational pull of some of the larger planets, Saturn and Jupiter in particular. The time at which they were ejected can be traced to a particular period when Jupiter and Saturn were in resonance. 'Resonance' here means that an exact number of orbits of the one planet equals a different exact number of orbits of the other; in this case two orbits of Jupiter corresponded to one of Saturn. The repetitive nature of the resulting arrangement meant that the successive tugs of the two large planets on the minor bodies produced a cumulative effect that probably ejected most comets into interstellar space. Only those few that for some reason did not get so many or such strong tugs, and did not quite make it into interstellar space, were left behind.

The nature of comets became progressively clearer during the late twentieth century, with great leaps forward in our understanding after a series of rendezvous by spacecraft with comets, starting in 1986 when the ESA's Giotto spacecraft flew close to Comet Halley. Comets are made of ice mixed with dust: the primitive material of the solar nebula. When one of these icy bodies falls in towards the Sun, it is progressively heated. On its surface, the ice fuses into a strong, dirty crust, like the snow left behind in piles in a city street as spring approaches. The action of sunlight on the dirt creates tarry substances that make comets as black as coal. These tars are organic chemicals – carbon-based, such as those found in a furnace or a fireplace – and on comets they are produced not by heat but by ultraviolet light from the Sun.

In 2001, the NASA Deep Space 1 spacecraft imaged the surface of Comet Borrelly with enough detail to see a very dark, soot-like material that hides the ice below. This picture was confirmed as typical in 2005, when the Deep Impact probe blasted a crater on Comet Tempel 1, releasing a cloud of water ice. From time to time, when an asteroid hits a comet, the same thing happens. But material can also be released from comets without a crash. Below the black, crusty, solid surface, the internal ice sometimes vaporizes with the heat of the Sun. Pressure builds up beneath the surface of the comet on its warm side and, from this steam kettle, the ice may explode out in a fountain of ice and dust fragments. This was observed not only on Comet Halley, by Giotto, but also on Comet Borrelly, by Deep Space 1, and on Comet Wild 2, viewed by the Stardust spacecraft in 2004. All of these bodies showed similar fountains of ejected material (10). As a result, a comet leaves behind a trail of dust and gas. The dust reflects sunlight; the gas absorbs sunlight and re-radiates the energy as radiation. The dust and gas form the comet's characteristic luminous tail (11).

The jets of steam, hail and rocks act like a rocket engine to give comets a push in the opposite direction. The forces that the jets create on comets make them erratic and even now, with today's advanced knowledge of celestial mechanics, and powerful computers, it is impossible to calculate the orbit of a comet to complete accuracy because of these unpredictable and arbitrary shoves off course.

It seems hard to imagine life developing and surviving on the unstable surface of a comet, but what is clear is that the tarry, smoky substances that cover comets include many complex chemicals. Some may be left over from the solar nebula that formed the Sun, but other chemicals are formed by the action of sunlight on the dust mixed in with the ice. These chemicals – termed 'prebiotic' because, although they are not biological, they can be used by life – are spread throughout the Solar System by the errant paths of disintegrating comets, delivered on occasion to a planet in bulk, when one crashes to its surface.

The impact of a comet with a planet was seen when Comet Shoemaker-Levy 9 was observed to pass close to Jupiter, split into a couple of dozen pieces, and then, on the next pass round the giant planet, crash into Jupiter's clouds, piece by piece, a few hours apart, over a week in July 1994. What might have been a small comet crashed into the Earth on 30 June 1908, exploding in the atmosphere over the Tunguska River in Siberia. The impact created an airburst explosion rated at 15 megatons, which flattened trees but left very little in the way of asteroid or meteorite fragments. Whatever the comet was made of – mainly water – was vaporized and distributed through the atmosphere. The water has fallen as rain all over the world; the last glass of water or wine that you drank contained drops of the Tunguska melted comet.

The asteroids are a mixed bunch of solid, rocky bodies. The biggest one in the Main Asteroid Belt – and the first asteroid to be discovered, in 1801 – is Ceres. It is 930 km in diameter, a third the size of our Moon. Ceres is a sphere because its own force of gravity overcomes the structural strength of the rocks it is made of, causing them to settle down into concentric layers. It is the most massive of all the asteroids. The next largest are Pallas and Vesta, each nearly spherical, and about 500 km in diameter, a sixth the size of the Moon. The bigger asteroids are, in a sense, 'failed planets'. When the planets were being made in the solar nebula, about 4,500 million years ago, Jupiter formed. It is such a massive body that its force of gravity stirred up the solar nebula just inside where it was orbiting. This prevented the partially built bodies forming here from completely coalescing. As a result, numerous nascent planets were created. It was a crowded part of the Solar System, and massive collisions were inevitable. Some of these colossal smashes broke their parent bodies into smithereens, creating a band of fragments; these bits and pieces of failed planets are the majority of the asteroids in the Main Belt at this time. They range downwards in size from Pallas and Vesta (**12**). Only thirty asteroids are more than 200 km in diameter, but there are about 750,000 bigger than 1 km in diameter and 25 million over 100 m. Presumably the pieces range down from metres to microns; the numbers must be uncountable.

When one of these small asteroids is destined to fall into the Earth's atmosphere it is termed a meteoroid. Friction of the meteoroid with the air causes a meteor trail to zip through the atmosphere, leaving a momentary, fiery trail of hot abraded dust, colloquially known as a 'shooting star'. If a residual piece of the original rock falls to the ground as a solid lump, it is then called a meteorite. In the only completely documented case, on 7 October 2008, a small asteroid/meteoroid called 2008 TC3 was tracked by an optical telescope on Mount Lennon in Arizona for about a day as it approached Earth. Typically, a meteoroid is the size of a grain of sand and its trail is as bright as a star, but this one was several metres in diameter. As it entered the atmosphere over Northeast Africa during the night it made a fireball as bright as the Moon that lit up the desert for hundreds of kilometres around. The fireball was observed by the pilots of an airliner 1,400 km away, as well as by meteorological and defence satellites inspecting the Earth from above. The meteoroid fell in Sudan as a meteorite, pieces of which were recovered two months later from the Nubian Desert (one small piece was presented to and is on display at the telescope where the asteroid was discovered). So meteorites are simply small asteroids that have fallen to Earth, in the process delivering whatever they are made of to our world.

* * *

In 1969 a meteorite broke up over Murchison, near Melbourne, Australia, and pieces fell across the town. There were many eyewitnesses (church-goers on a Sunday morning), who promptly collected about 80 kg of fragments. In a blanket search of some of the fragments, the meteorite was found to contain tens of thousands of complex organic chemical groupings, which could be components of millions of organic molecules. In targeted searches for particular molecules, researchers found many amino acids – a type of complex compound essential for life because it builds up the proteins that make biochemistry work. Indeed, the Murchison meteorite contained many but not all of the twenty amino acids that terrestrial life uses, as well as about fifty others. Because they were found deep inside the fragments, and there had been so little time for terrestrial compounds to have infused into the meteorite material, it is generally agreed (but not quite unanimously) that these substances are not terrestrial contaminants: they were made in space.

This conclusion is strengthened by other variations between the amino acids in the Murchison meteorite and those that are typically found on Earth. One significant difference is the 'chirality' of the molecules. Chirality is the 'handedness' of the molecules. Most amino acids are chiral, that is to say they exist in both left-handed and right-handed forms. What does this mean? Position your hands as if you are praying in the modern Christian manner, with palms pressed together. Everything in your hands matches – your thumbs line up, and so do your fingers of different sizes, your knuckles, your wrists and so on: each hand looks the same. But if, instead, you hold your hands out in front of you, palms up, it is clear that each one is different; you could not superimpose your left hand onto your right hand. Your hands are mirror images, and the scientific word for this is 'chiral'. Amino acids are chiral molecules.

If, in a laboratory, a chemist builds up an amino acid from simple non-chiral components, the amino acid is produced in equal numbers of left- and right-handed forms; the scientific terminology calls this mixture 'racemic'. The amino acid molecules used by life on Earth are (with some exceptions) left-handed only. Scientists have never, through chemical methods alone, been able to start with simple non-chiral components and build purely left-handed amino acids; but nature does, and it also happens in space. The Murchison meteorite is not racemic. It is also not entirely left-handed, but left-handed forms do predominate. It is possible that the amino acids in the Murchison meteorite were racemic when they arrived from space, but were contaminated by left-handed amino acids produced by terrestrial life. The meteorite may not be as clean and uncorrupted as people think. It is also possible that the amino acids in meteorites have been pushed to be more left-handed than right-handed, and there is a cosmic process that is the reason why terrestrial life developed to use left-handed amino acids.[1]

All this is provocative but controversial. The more definite implication is that basic biochemicals are made by natural processes throughout the Universe, perhaps, as we have seen, on the surface of comets. Somehow these biochemicals got into the meteorite – it had been orbiting in the Solar System for a long time, and had many opportunities to pass through the tails of comets and pick up dust – or perhaps the biochemicals had been made in the material of which the meteoroid was formed originally. As the Murchison meteorite showed, chemicals can move from place to place in the Solar System, possibly even from further afield – from interstellar space perhaps. This makes astronomers think that complex molecules can be made in one place, and seed life somewhere else; for example, on the surface of another planet.

* * *

The largest asteroids are spherical, and are dignified by the name 'dwarf planets'. The smaller an asteroid, the more likely it is to be irregular in shape. Astronomers first got clues that some asteroids were non-spherical when they saw that some varied in brightness as they rotated, presenting alternately smaller and larger faces towards Earth. Asteroids are so small, however, that it is impossible to image them from Earth to see what shape they really are. The Hubble Space Telescope has a sharper view, from its position in the clarity of space, but even this telescope has produced pictures only of the surface of the largest asteroids: Ceres, which is very spherical, and Vesta, which is not. Vesta has a gigantic piece missing at its south pole (12). Using radar techniques, radio telescopes have had some success in producing images of a few asteroids, the orbits of which happen recently to have taken them very close to Earth: those called Casalia, Geographos and Toutatis are three good examples. Only with spacecraft flying past, or even deliberately rendezvousing with, an asteroid, can its true shape be seen clearly. All, save the very largest, resemble potatoes.

The modern interest in asteroids and their variety is reflected in the number of spacecraft that have been programmed to fly by, or sent to rendezvous with, asteroids in the past twenty years. The first asteroids to be imaged were Gaspra and Ida, viewed in 1991 and 1993, respectively, by the Galileo probe as it was travelling to Jupiter. Ida proved to be two asteroids; it has a small moon, Dactyl. Three more spacecraft flew past asteroids on their way to comets: Deep Space 1 visited the asteroid Braille in 1999; Stardust visited Annefrank in 2002; and Rosetta visited Šteins in 2008 and Lutetia in 2010. The first probe sent specifically to an asteroid was NEAR Shoemaker, which videoed Mathilde in 1997, and went on to orbit around Eros, landing there in 2001. In 2005, the Japanese Hayabusa ('peregrine falcon') probe investigated Itokawa, landing on it and even returning samples of its surface dust to Earth. NASA's Dawn mission arrived in July 2011 to orbit Vesta for a year, and will go on to Ceres in 2015.

All the asteroids investigated are covered with impact craters, caused by collisions with meteoroids or small asteroids. Collisions between asteroids are common. In January 2010 the Lincoln Near-Earth Research (LINEAR) Program Sky Survey spotted a curious asteroid with a comet-like tail. A detailed picture by the Hubble Space Telescope showed a bizarre X-shaped object at the head of the comet-like trail of material. This proved to be the slowly expanding dust cloud around a collision between two asteroids that occurred in the Main Belt in February or March 2009. The dusty cloud around the new object, P/2010 A2, which is an asteroid some 120 m in size, is probably the result of a collision with a smaller rock, perhaps 3 to 5 m wide. The collision speed was probably about 18,000 km an hour, releasing about the same energy as a small atomic bomb. The encounter likely made a crater about 20 to 30 m in diameter in the larger asteroid. Although this is the first such event to be observed, modest-sized asteroids smash into each other roughly once a year. These collisions spread asteroid material throughout the Solar System, and whatever chemical materials they contain are widely distributed.

Should a planet pass near to a cloud of asteroid fragments, it would experience a sudden meteor shower: a bombardment of meteor fragments. There is evidence for such an event happening to the Earth in the Ordovician geological period, approximately 470 million years ago (see **Table 1**, p. 69). Ordovician limestone rocks found in Scandinavia contain fossil meteorites. The first was discovered in 1952 by stone workers in a quarry. It was about 10 cm in size, and languished in a geological collection until 1979, when its nature was realized. Now ninety similar fossil meteorites are recognized: all of them are of the same type, and all of them fell to Earth at about the same time. In the same rocks, in the same layers as the recognizable lumps of fossil rock that are meteorites, are tiny bits of original meteorite material in the form of little dust grains. At the other extreme of size, there are at least four meteor craters – the age of which matches the right time span – possibly caused by large fragments of the same shower. Even today, 20 per cent of meteorite falls are of the same type of rock as the Ordovician meteoroid fall, resulting from the same asteroid collision. Presumably some of the fragments from the collision that created all these meteorites that have fallen to Earth are still in orbit in the Solar System. It is possible to get some idea of the composition of the rock type that any given asteroid is made of by looking at the distribution of sunlight that it reflects. Just as the colour of a rock indicates to a geologist what minerals it is made of, so the colour of an asteroid shows its composition. Astronomers measure colour in an astronomical object by dispersing its light into a spectrum, the colours of the rainbow from blue to red, and determining the balance of light in these component colours. The spectrum of the asteroid Gefion matches the composition of the Ordovician meteorites. It is the largest of a number of asteroids that have the same orbit, called the Gefion family, all created in the last quarter of the history of the Earth, in what Swedish geologist Birger Schmitz (b. 1955) described as the 'biggest bang in the Solar System for a billion years'.

Gefion is probably covered with the dust generated in the asteroid collision. There are plenty of indications that asteroids are covered with dust made in this way. The close-up views of Eros from the NEAR Shoemaker spacecraft showed dust and pebbles across the whole surface of the asteroid, and when the Hayabusa spacecraft landed on Itokawa, it kicked up dust that it returned to Earth. As expected, the dust had a similar composition to that from meteorites. Dust from asteroid collisions permeates the Solar System throughout its orbital plane, the 'ecliptic'. Asteroid dust is the medium that reflects sunlight in the phenomenon known as the 'zodiacal light', a cone of light that shines up from the horizon after sunset, its axis along the line of the ecliptic. In dark skies, it shines like a searchlight beam along the zodiacal constellations. Our Earth is bathed in asteroid dust.

All this seems to argue against any possibility that water might be found on asteroids; the surface environment of dry, dusty asteroids, warmed by collisions, does not at first seem to be somewhere that one might find water or life. Contrary to this expectation, however, a few asteroids do show faint tails, like comets. Given that the tail of a comet is dusty material released when the ice of which it is composed melts, this has made astronomers think that some asteroids have water on their surfaces. More direct evidence for this was discovered in 2010, by two groups of astronomers using the NASA Infrared Telescope Facility (IRTF) on Mauna Kea in Hawaii. They observed the spectrum of the Main Belt asteroid Themis and inferred that it is covered with a frosty coating, containing ice and organic chemicals, somewhat like the surface of a comet.

Themis is only about 200 km in diameter and has no atmosphere, so the presence of surface ice is surprising. If the ice was put there 4,500 million years ago in the solar nebula, and the asteroid has been in the Main Belt for this long, warmed gently by the Sun, one might have expected that the ice would have sublimed and evaporated long ago. It is possible that the asteroid is an interloper to the Main Belt, having migrated there from more distant, colder regions, where ice has persisted. We could also imagine that ice may be held in reservoirs beneath the asteroid's surface, shielded from the Sun, and that it is slowly churned up as the soil of the asteroid is 'gardened' by the collision of smaller asteroids. This could replenish the surface ice. It is also possible that it was delivered to the asteroid recently, for example if it was struck by a comet that melted on impact.

* * *

All eight planets – except Mercury and Venus – have one or more moons or natural satellites that orbit around them. What are the prospects of finding life, or the chemicals that make life, on them? Like the asteroids, the moons are a mixed bunch, so the answer to that question varies from moon to moon. The largest satellite is Ganymede, orbiting around Jupiter, with Titan, orbiting Saturn, running a close second. Both are more than 5,000 km in diameter, a little bigger than the planet Mercury, and considerably larger than our own Moon, at 3,450 km. Saturn has a myriad of tiny moons, centimetres in diameter (and perhaps even smaller), each in an individual orbit within a disc that makes up the planet's ring system. There are many satellites orbiting the planets that range in all sizes in between. Jupiter has four main satellites, all of comparable size, which revolve in a single plane and are related by birth; it appears that they were made in an eddy of the solar nebula at the time that Jupiter was forming. Other satellites are asteroids that have been captured by the planet: they ventured very close and were brought into orbit around the planet by the attraction of its gravity. The planet Mars has two small satellites, Phobos and Deimos, which are the same potato shape as typical asteroids, although it is a mystery why their orbits are so well aligned with the equator of Mars. Typically, captured satellites orbit a planet in such a way that the plane of the orbit lies at a large angle to its equator. The orbit is inclined at an angle that 'remembers' how the asteroid first approached the planet. The encounter can be from any direction, so the orbit typically bears no relationship to the spin of the planet and the plane of its equator. In a few extreme cases, the encounter was in the opposite sense to the orbital motion of the planet and its rotation, and these satellites orbit backwards. This is so with the moon Triton around Neptune. It is strange, therefore, that Phobos and Deimos seem to have approached Mars so as to line up with its equator. But if they are captured asteroids, what has already been said about life on asteroids applies to these moons.

All the natural satellites are solid bodies, as is Earth, and a few, generally the largest, have measurable atmospheres: Io, Callisto, Europa, Ganymede (Jupiter), especially Titan (Saturn), Titania (Uranus) and Triton (Neptune). They are moons, subordinate to their planet because they are in orbit around it. But in themselves they are more like worlds in their own right. Ice is common on the surface of the satellites that orbit the outer planets. The surface of Europa (13) is in fact entirely covered by ice, and cracks show where the ice floes shift and grind against one another. According to theoretical cal-culations, the depth of the ice is considerable: from a few km to 30 km thick. Below the ice is a large amount of liquid water, perhaps as deep as 100 km. The water is liquid, even though Jupiter lies well outside the habitable zone. The reason is that the ice is warmed from below by radioactivity, and by tidal heating caused by the flexure of the body of Europa caused by Jupiter's force of gravity. The thick, permanent ice cover, far from the warmth of the distant Sun, prevents evaporation of the liquid water and maintains the ocean below. The total amount of water on Europa is possibly greater than the amount on Earth. Given the importance of the oceans in the story of the development of life on our planet (chapter 7), it is clear that in our search for life elsewhere we must return to study Europa more closely. I will do this in chapter 14. In fact, there may be liquid oceans inside several of the icy moons or other small bodies of the distant Solar System, and presumably of other planetary systems. This environment may be, by volume, the most abundant habitable place in the Universe.

Finally, there is our own Moon. It is next to the Earth and therefore in the habitable zone. Even with the unaided eye, one can distinguish grey patches on the surface of the Moon, optimistically called *maria*, Latin for 'seas'. Nevertheless, the Moon is mostly a dry world, covered with a layer of dust that takes deep impressions of astronauts' boots. The Apollo missions visited only the region of the lunar equator, however; there is in fact ice at the Moon's poles. It exists in craters, eternally shadowed by high ramparts (**14**). This ice was probably brought to the Moon by the impact of a comet. The Moon would temporarily have had an atmosphere of melted water. The water vapour mostly dissipated into space, lost from the Moon because of its weak gravity, but some froze on the coldest parts of the surface, in the shadows. What we know for certain, because astronauts have been to the Moon during the Apollo programme, is that most of the Moon is dry, dusty and sterile. The lunar soil brought back for analysis contains no dead bodies. The Moon is devoid of life, even of defunct remains.

I started this chapter with the assertion that all the life that we know of exists on Earth. What this quick tour of our planetary system shows is that it is more accurate to say that life lives in the Solar System. Water and biochemicals have been, and still are, fed to Earth from distant regions and sporadic impacts. Life has flourished here on Earth, which is undoubtedly the planet that has it in the most abundant and obvious forms. But the potential for life is ubiquitous in our Solar System, and there are niche environments, if not whole planets, where it may have established a foothold. Life is a phenomenon that is characteristic of an entire planetary system.

5

The Earth, Life's Home

What we know for sure about life in the Cosmos is that it exists here on Earth. Inevitably, and as we have seen in chapter 4, this terrestrial case, with modest extrapolations, guides what we think are the essential ingredients for life. There is an obvious limitation to this approach, and we must bear in mind that there may be other models for life. But we can never hope to understand extraterrestrial life if we do not understand what is here on Earth. From what we do know, life emerged naturally from carbon-based molecules on our planet about 4,000 million years ago, in conditions that could be common throughout the Universe. This gives us hope that we might find extraterrestrial life, if we can only find the right places to look.

The history of life on Earth is linked through the investigation of its vestigial remains in successive layers in the ground. The surface of the Earth is subject to weathering: that is, when the ground is worn away by volcanic action, wind, ice or water, and redeposited in drifting ash, mudslides, sand dunes and, in particular, in the muddy or sandy sediment at the bottom of lakes and seas. As a result, the remains of dead forms of life, or such traces as footprints or burrows, can be covered gently without much disturbing their form. The remains themselves may decay to gases or soluble salts that dissipate, but other chemicals can replace those that have been lost – in the process creating a stony replica of the original remains. This imprint, of a formerly living creature in rocks, is called a fossil (from the Latin word meaning 'dug up'). The earth thus contains a record of the forms of life throughout time.

The probability that a life-form produces a fossil is small. Vegetation rots rapidly, unless it drops into acidic water where it might be pickled, or sinks into a marsh or to the bottom of a pond, where it might be silted over. The corpse of a dead animal might be scavenged by carrion eaters, its soft tissue might completely decay by the action of fungi and bacteria, the body might be scattered by the wind or trampled by animals, abraded by sandstorms or dehydrated by the Sun, weathered by rain, or crushed and ripped up by a violent natural burial process, such as a landslide. Land creatures in particular are less likely to become fossils, because they live in an environment that is readily open to intrusion. By contrast, marine creatures are enclosed by water and may be buried in fine-grained, slowly falling sediment. Marine fossils are more common. The rocks themselves are in flux, so even if a fossil of a creature is made, it may later be destroyed. As a result, the fossil record is full of gaps.

Fossils are exposed as rocks come to the surface and weather away. This begins as tectonic plates collide, their crushing action producing wrinkles on a scale that raises mountains. When the rocks at the summits are weathered, they are redistributed in low-lying areas, including under water. The deposited material lies on top of older rocks. Its weight adds pressure onto the material below, compacting it and fusing it into solid forms, even, through a process of metamorphosis, changing its chemical or crystalline structure. Thus rocks are built into strata, with, in principle, the older rocks below, the more recent above and the most recent at the surface of the planet. Later tectonic activity may disturb the sequences of strata, folding the regular, flat layers into waves, splitting strata vertically in planes and thrusting one side of the fault line higher than its neighbours, even reversing the order of the strata.

* * *

Rock strata may become visible at cliff faces. If, by tectonic action or by the cutting effect of water, a vertical section is cut through the ground, its layers may be revealed. Of course, if the cutting action is violent, the layers may be jumbled. Thus, layers of rocks in some areas of the Earth are very difficult to sequence. Other areas of the planet have remained less disturbed, and the sequencing is clearer, as in the walls of the Grand Canyon, in northern Arizona (**15**). Putting together all the evidence about the sequencing of strata around the world, including the fossils that they bear, geologists have compiled variations of **Table 1**, classifying rocks according to their character and age. **Table 1** is the Earth's curriculum vitae, the timeline of its development.

The rim of the Grand Canyon is on a high, flat, limestone plain, known as the Colorado plateau, laid down about 250 million years ago at the bottom of a shallow sea. There are younger, softer rocks that overlie the limestone further north. The softness of the rocks in that region is readily visible in the dramatically eroded landscape of Bryce Canyon in southern Utah; in the Grand Canyon area, however, this layer has been completely washed away. Below the topmost limestone layer lie deeper and deeper horizontal sedimentary layers of sandstone, shales and mudstones, which represent the successive action of lakes,

Table 1. The Geological Timescale

Based on the standard table of the International Union of Geological Sciences: International Commission on Stratigraphy.

Eon	Era	Period	Ended (million years ago)	Began (million years ago)	Age of Earth (millions of years)	
Phanerozoic	Cenozioc	Neogene	0	23	4,577–4,600	
		Palaeogene	23	65	4,535–4,577	
	Mesozoic	Cretaceous	65	145	4,455–4,535	Chicxulub meteorite impact
		Jurassic	145	200	4,400–4,455	Breakup of Pangaea Era of the dinosaurs
		Triassic	200	251	4,349–4,400	
	Palaeozoic	Permian	251	299	4,301–4,349	Formation of Pangaea
		Carboniferous	299	359	4,241–4,301	
		Devonian	359	416	4,184–4,241	The first animals
		Silurian	416	444	4,156–4,184	
		Ordovician	444	488	4,112–4,156	
		Cambrian	488	542	4,058–4,112	Cambrian explosion of life
Proterozoic	Neoproterozoic	Ediacaran	542	650	3,950–4,058	
		Cryogenian	650	850	3,750–3,950	
		Tonian	850	1,000	3,600–3,750	
	Mesoproterozoic	Stenian	1,000	1,200	3,400–3,600	
		Ecstasian	1,200	1,400	3,200–3,400	
		Calymmian	1,400	1,600	3,000–3,200	
	Palaeoproterozoic	Statherian	1,600	1,800	2,800–3,000	
		Orosirian	1,800	2,050	2,550–2,800	
		Rhyacian	2,050	2,300	2,300–2,550	Oxygenation of Earth's atmosphere
		Siderian	2,300	2,500	2,100–2,300	
Archaean	Neoarchaean		2,500	2,800	1,800–2,100	First fossils
	Mesoarchaean		2,800	3,200	1,400–1,800	
	Palaeoarchaean		3,200	3,600	1,000–1,400	
	Eoarchaean		3,600	3,800	800–1,000	Beginning of life on Earth? Formation of the oceans
Hadean			3,800		500	The Late Heavy Bombardment
					50	Oldest known rocks Formation of Moon
				4,530	0	Formation of Earth

Note to Table 1. The geological life of the Earth is divided into eons, each roughly a billion years long, subdivided into eras, which are themselves subdivided into periods. In each of these timespans the rocks have similar characteristics, in terms of the circumstances in which they were laid down and the fossils they contain. The Paleogene period is also called the Tertiary.

estuaries and seas in the area, as the climate and sea-level changed over a period of, in total, 570 million years.

About halfway down the walls of the canyon is the 'Great Unconformity'. The strata here are no longer horizontal, but have been folded and altered in angle by the pressure of up-swelling volcanic rocks below, in events dating to about 800 million years ago. The upper section of the folded hills and mountains that were created at that time has been eroded in a horizontal plane, which became the bottom of the succession of seas that created the rocks of the upper levels, above the Great Unconformity. Between the volcanic events that distorted the lower layers 800 million years ago, and the start of the regular layering that created the upper layers 570 million years ago, about 230 million years of deposits have gone missing in the erosion process. Below the Great Unconformity there are a further 420 million years of sedimentary rocks, which date back to 1,450 million years ago. At the very bottom of the canyon, at the level of the Colorado River, are hard volcanic rocks dating back 1,700 million years. They are the roots of a huge mountain range that formed in this area, all of it eroded away and redeposited in the layers of the canyon above. It is as if the present-day Rocky Mountains have been sanded off and scattered over the western half of North America. The layers have been carved away and exposed to view by a combination of the action of the Colorado River and the uplift of the Colorado plateau during the last 25 million years.

Similar sequencing of rocks can be recognized across the world. The rocks that lie just above the Great Unconformity are so-called Cambrian rocks, which are of a similar age to those found in Wales, in the western part of Great Britain, by the nineteenth-century geologist Adam Sedgwick (1785–1873) (Cambria is the Latin name for Wales). The Unconformity represents a huge change in the global climate that occurred 542 million years ago. Since that time, life flourished here on Earth. The event that changed the climate and triggered this biological surge has become known as the Cambrian explosion.

In the Cambrian strata of the Grand Canyon (laid down from 542 to 488 million years ago), and in other rocks of the same age from around the world, geologists have found abundant fossils – but only in rocks deposited at the bottom of the sea. Apparently, only marine life existed then on Earth; the dry land was devoid of creatures. The earliest of the Cambrian rocks contain fossils of sea creatures, such as trilobites. Trilobites are now extinct, but they were one of the most long-lasting kinds of animal ever to exist, persisting for 270 million years. They varied in size – from about a millimetre to as big as a dinner plate – and had a hard outer skeleton over their backs, out from under which stuck unprotected legs. The creature of today that most resembles the trilobite is the horseshoe crab. Some species of trilobites swam in the sea, while others ranged over the seabed. The different features of the trilobite fossil species are well studied, and this makes them valuable to geologists as dating markers for otherwise problematic rock layers.

In the layers above the Cambrian rocks of the Grand Canyon, and elsewhere, are more diverse fossils. The layers below the Cambrian rocks are referred to as pre-Cambian. Naturalist David Attenborough (b. 1926) recalls that as a schoolboy in Britain, interested in fossils, he was told not to bother looking for them in the older layers, as life appeared on Earth only at the time that the Cambrian rocks were laid down. This has turned out not to be true. It was proved wrong in 1957, by another British schoolboy, Roger Mason (b. 1940), now an Earth sciences professor at the China University of Geosciences at Wuhan, who found a frond-like fossil of a life-form called charnia in rocks in a quarry in Leicestershire, central England, that were undoubtedly pre-Cambrian. This discovery opened the eyes of palaeontologists to the possibility that other pre-Cambrian fossils exist. Today the list of rocks in which fossils have been found stretches back to those more than 3,000 million years old. Life must be at least this old, not 'only' existing in the 550 million years since the Cambrian period. But before describing the earliest life on Earth, we should look at how we know the ages of rocks, indeed of the Earth itself.

* * *

In 1650, the Protestant Archbishop James Ussher (1581–1656), of Armagh in Ireland, published an estimate of the date of the biblical Creation. Like many absurdities of scholarship, the result arose from pushing a principle beyond its limitations. The principle was that the Bible could, indeed should, be subjected to literal analysis.

Ussher used the genealogies of Old Testament figures as the core of a timeline that stretched forwards to his own day, and backwards to the events of Creation in the first chapter of the Book of Genesis. The biblical text of the Christian Old Testament, deriving from the Written Torah or 'Tanakh' of Judaism, runs directly from the Creation of the Universe and the Earth to the appearance of the first two humans, Adam and Eve, and then describes an unbroken sequence of their descendants, the Prophets and others, up to the events of the Christian New Testament gospels. The gospels mention various matters of Roman history and can be directly linked to the modern calendar.

Ussher's chronology of events was reproduced in the standard edition of the Christian Bible, which was used in England for centuries, and became famous and accepted. He set the date of the creation of the Universe, the Earth and living creatures at 4004 BC. This conclusion is still accepted by those Christians who follow the strict principles of fundamentalism on which Ussher based his work: the modern-day Creationists.

But in the nineteenth century, British geologists, including Charles Lyell (1797–1875) and John Phillips (1800–1874), began talking in terms of millions of years for the Earth's age. They based their calculations on how long it would take for sedimentary rocks to be laid down from sea deposits or for rocks to be eroded away. The discrepancy between the scientific and the biblical age of the Earth was made worse by the calculations of Lord Kelvin (1824–1907), a physicist who suggested that hundreds of millions of years was nearer the truth, and the issue was further exacerbated at the turn of the nineteenth and twentieth centuries, when the French scientists Antoine Henri Becquerel (1852–1908) and Marie Curie (1867–1934) and Pierre Curie (1859–1906) discovered radioactivity. The British physicist Lord Rutherford (1871–1937) had the idea of applying this phenomenon to date the decay of radium in rocks, a process known as radiometric dating. Radium produces helium when it decays, and if the radium is embedded in the right kind of rock the helium is trapped and accumulates over time, without loss. The less radium and the more helium there are, the older the rock is. By 1907, a young American chemist, Bertram B. Boltwood (1870–1927), had discovered that some rocks were, by his measurements and calculations, as much as 1,000 to 2,000 million years old. By the middle of the twentieth century, geologist Arthur Holmes (1890–1965) concluded – from the radiometric dating of old rocks all over the world – that the Earth was at least 3,000 million years old.

The way radiometric dating works is as follows. An isotope is an atomic nucleus constructed in a particular way, with a defined number of protons and neutrons. Some of the constructions are unstable, and these unstable isotopes are liable to decay to other isotopes, creating what is described as a parent–daughter relationship. This is what the phenomenon of radioactivity is. Modern methods of radiometric dating are based on the decay of a range of long-lived radioactive isotopes that occur naturally in different rocks and minerals: see **Table 2**. The parent isotope decays to a daughter at a fixed rate, expressed as a half-life. This is the time it takes for half the amount of the parent isotope to decay. Uranium-238 has a half-life of 4,470 million years, so it is a particularly suitable isotope with which to measure the age of the Earth, which, as we will see, is 4,530 million years old. Because the half-life of uranium-238 and the age of the Earth are nearly the same, we can say that the Earth contains half the uranium-238 that it did originally. Not all methods work for all rocks. For example, if a rock does not contain uranium, or the daughter isotope is produced by another decay route, the measurements can be exquisitely precise, but the given method might not be accurate. The geologist has to select the right tools to measure the age of the rock in hand, and there may not be a tool that suits the rock and its circumstances.

Table 2. The Principal Isotopes Used in Radiometric Dating

Parent isotope	Daughter product	Half-life (years)
carbon-14 (^{14}C)	nitrogen-14 (^{14}N)	5,730
uranium-235 (^{235}U)	lead-207 (^{207}Pb)	704 million
potassium-40 (^{40}K)	argon-40 (^{40}Ar)	1,250 million
uranium-238 (^{238}U)	lead-206 (^{206}Pb)	4,470 million
thorium-232 (^{232}Th)	lead-208 (^{208}Pb)	14,000 million
lutetium-176 (^{176}Lu)	hafnium-176 (^{176}Hf)	35,000 million
rhenium-187 (^{187}Re)	osmium-187 (^{187}Os)	43,000 million
rubidium-87 (^{87}Rb)	strontium-87 (^{87}Sr)	48,800 million
samarium-147 (^{147}Sm)	neodymium-143 (^{143}Nd)	106,000 million

In **Table 2**, the number beside the name or symbol of the atom denotes the number of protons and neutrons in the atom's nucleus, and designates the particular radiometric isotope. The first pair of isotopes is used primarily for dating dead organic material, such as wood, over a matter of a few thousand years at most. The remaining pairs are used for dating old rocks; the three principal pairs used in modern geology are K–Ar, Rb–Sr and U–Pb.

The earliest rocks on Earth that have been dated this way are found in the Canadian Shield (the geological formations of Canada, north of the Great Lakes, between the St Lawrence River and the city of Yellowknife), Australia and Africa, and are generally between 2,500 and 3,800 million years old. The oldest are 4,031 ± 3 million years old, rocks known as the Acasta Gneiss of the Slave craton[1] in the Northwest Territory of Canada. Rocks between 3,800 and 4,280 million years old have been found in the Nuvvuagittuq greenstone belt on the coast of Hudson Bay, in northern Quebec. The very earliest material that has been found on Earth is older, but has been weathered and recycled by geological processes into younger rocks. This material consists of a number of zircon crystal grains, aged an average of 4,350 ± 50 million years, which are part of a sandstone material from the Jack Hills of the Narryer Gneiss Terrane[2] of Western Australia. The sandstone was laid down about 3,060 million years ago.

The limitation of Rutherford's method is that so much has happened to the rocks of the Earth that it is difficult to find samples that conform to the ideal prescription, in which the daughter products have not escaped from the rock and altered the proportion with their parents. Volcanic material has welled up from the liquid interior of our planet to overlay the old, and tectonic plates have collided together, scrambling old and new material. Rocks have weathered and re-compacted. It is hard to find rocks that have remained undisturbed since they first solidified; these are exposed on the Earth's surface in very few places.

The figure that scientists accept for the date of the formation of the Earth is now based on material brought back from the Moon by the Apollo astronauts, and from meteorites that have fallen to Earth from space, following the assumption that all of the planets and planetary material in the Solar System, including the Earth, formed at the same time.

The Moon has not been active with volcanoes for a long time. Its surface layers have been gardened by the impact of meteoroids and asteroids, which made the lunar craters. These impacts both melted surface material and excavated subsurface material onto the surface. This changed some of the original composition of the surface rocks, but also exposed rock fragments that have been preserved intact since the formation of the Moon. In 1971, when Apollo 15 landed on the Moon – in Mare Imbrium, at the base of the lunar Apennine Mountains – two astronauts, David Scott (b. 1932, the mission commander) and Jim Irwin (1930–1991, the lunar module pilot), foraged in moon-dust near the rim of Spur Crater, and Scott spotted a rock that took his fancy (16). It had an unusually coloured piece embedded within it. Scott picked it up with long-handled tongs. The two men looked at it, shaking the dust off. The rock glinted, and they knew immediately it was crystalline. Preserving it in a sample bag, they carried it back to Earth. This rock, which came to be known as the Genesis Rock, proved from its radioactive properties to be 4,500 million years old.

Another source of information about the age of the Earth is meteorites. Astronomers believe that some meteorites consist of material that has been left over from the time of the formation of the Solar System, material from the 'solar nebula': a flat disc of stellar material that pancaked into orbit around the Sun as it contracted from a near-spherical interstellar cloud of gas and dust. Although most meteorites are recently broken fragments of asteroids and comets, some meteoroids may have been in orbit in space from the earliest times, undisturbed until they encountered Earth and fell to ground. The ages of the oldest meteorites, as measured by the properties of their radioactivity, range from 4,530 to 4,580 million years old. It is astonishing to think that, even at a distance in time of 4,500 million years, the spread of ages of 50 million years is not an inaccuracy or vagary in the measurements. The difference in ages of these primitive meteorites is real, and is due to the period of time that it took the solar nebula to build up significant lumps of rock from grains of dust into meteoroid-sized masses.

From the perspective of life on Earth, our planet has performed well. It has provided a stable enough platform for the development of life over a long period of time: 4,500 million years. There has been a gradual evolution in the character of the Earth from eon to eon, as shown in **Table 1** (p. 69). There have been changes in global climate, marked by non-conformities in the eras of **Table 1** and in the fossil content of rocks. For nearly all this time, however, Mother Earth has patiently fostered the development of life from its early beginnings, as we shall see in the next chapter. When we search for life elsewhere in the Universe we will be looking for a similar environment that has lasted a comparable length of time.

6

Life Gets Going

The earliest forms of life on Earth are those closest in nature to the prebiotic chemicals that are made everywhere in the Cosmos, elaborated near stars and delivered by comets and asteroids from space to our planet. Since these processes are universal, it seems reasonable to expect the same starting point for life on every planet, even if thereafter evolution takes it in another direction. Assuming a common prebiotic ancestor for all life in the Universe, determined by chemistry, primitive life on other planets could therefore be the same as the earliest forms of life on Earth. These forms are known as archaea, single-celled organisms similar to bacteria, so simple that their cells have no nucleus.

For 80 per cent of the history of the Earth so far, these single-celled organisms were the only forms of life. Their physical traces are not easy to identify. In the more recent past, there were abundant forms of life with large, hard parts, shells and exoskeletons. Rocks, often sedimentary, silted gently over the dead remains and preserved them as fossil shapes. The earliest forms of life on Earth, however, were small and soft-bodied, and generally the geological processes that laid down rocks were more violent, sometimes volcanic. Often the older rocks were disturbed or weathered into new material. Under such conditions, fossils were far less likely to survive.

There are some fossils that date back as much as 80 per cent of the age of the Earth, however, and these include 'stromatolites'. Stromatolites are layered rocks formed by the growth of blue-green algae and other micro-organisms, which trap small grains of material (such as bits of sand or limestone) and bind them into a succession of horizontal mats. A stack of mats grows up into a column that sticks up from the sea floor. The largest and most diverse collection of present-day stromatolites exists at Hamelin Pool, in the south of Shark Bay in Western Australia. Some living stromatolites are 3,000 years old, but their ancestors date back a million times longer.

Some stromatolites are fragile, but others are stable and covered by water in which small particles of sand are suspended. It is these stable forms that are in an ideal situation to become fossils. They are recognizable by the laminations of the rock of which they are formed, and the repetitive scale of the columns and pillars that are fused together.

The micro-organisms themselves are small and soft, and do not readily survive as fossils. Therefore, the stromatolite fossils may not be the remains of a life-form itself, but of the laminated structure that it built or of the circular cross-sections of the columns (17). If fossils of the micro-organisms themselves do survive, the signs they leave behind are difficult to recognize and distinguish with certainty from purely mineral, inorganic structures, such as crystals and similar growths. Moreover, laminated rock can be formed in many ways, most of which are nothing to do with mats of blue-green algae. Identification of some of the ancient stromatolite fossils is thus controversial.

The principal micro-organisms that form stromatolites today were previously termed blue-green algae. The name is now thought to be inappropriate for these creatures, which are made of single cells without a nucleus (a characteristic of bacteria), so such micro-organisms are now called 'cyanobacteria'. As their previous name implies, they are blue-green, and, in a mass drifting in the sea, a lake or a pond, they can be seen as a diffuse blue-green 'bloom'. In warm weather, when the population of cyanobacteria may explode, the bloom is sometimes washed up on shore in large quantities as a tide of green slime. The slime is mucus exuded by the cyanobacteria. It is the mucus that, in the right circumstances, may catch drifting sand and other debris, and cement the grit into solid mats. The mats may anchor to the seabed and grow, layer by sticky layer, into columns.

Each mat is a complete ecosystem, based on micro-organisms of one or more species, including cyanobacteria. Within a mat, the positions of species may vary as regards being optimal in terms of such elements as the flow of nutrients, light, water acidity and the like. For example, the bacteria that inhabit the underside of a mat (in its centre, where it is covered from both above and below) will be protected, but will also be washed by less food-carrying water than the cyanobacteria that live on the fringes of the mat.

As a result, a stromatolite column has some of the characteristics of a complex creature: cells, the function of which depends on position, within a physical structure. The various communities in a stromatolite make up complex associations of diverse species and create the structures that can be seen in fossils. With time, these communities diversify or specialize and become interdependent, with the total structure then able to evolve to suit different or changing environments. Each individual stromatolite, however, is so variable in the proportions and relative structure of cell communities that 'a stromatolite' cannot be thought of as a species in its own right. The bacteria have a similar relationship to a stromatolite as do bees to a beehive.

Cyanobacteria use water, carbon dioxide and sunlight to photosynthesize their food, through the mediation of the green pigment, chlorophyll. As present-day plants and trees do, they take in carbon dioxide and give out oxygen; currently, cyanobacteria are responsible for producing about a quarter of the world's oxygen. The cyanobacteria of pre-Cambrian times may have created the conditions for their own decline by producing oxygen in such abundance that they changed the composition of the Earth's atmosphere, thus making it possible for oxygen-breathing organisms to evolve and feed on cyanobacteria. Supporting this interpretation is the fact that a change of atmosphere took place at about the same time as the epoch of the earliest undisputed fossils of stromatolites.

There are more than 250 pre-Cambrian stromatolite deposits known worldwide, in countries that include Australia, the Bahamas, Brazil, Canada, China, France, India, Israel, Kazakhstan, Norway, Russia, South Africa (see 17) and the USA. The dates of the rocks in which the stromatolite fossils are found extend back at least 2,600 million years. The oldest claimed stromatolite fossils with microstructures resembling bacteria come from 3,430-million-year-old rocks called cherts, in Pilbarra, Western Australia. Counterclaims suggest that these supposed oldest examples of life are nothing more than tiny cracks that are packed with minerals, such as haematite (an iron oxide), which accidentally take the form of bacteria as the mineral seeps into the rock. There is widespread agreement that the oldest undisputed microfossils come from rocks that are about 2,600 million years old; any earlier examples are controversial.

*   *   *

Even more uncertain is the evidence for the life-forms that existed before cyanobacteria, based on inferences about the evolutionary relationship of the bacteria to other forms of life. The idea that there is a continuous spectrum of life-forms originated with medieval theologians, when they envisaged a great 'chain of being' that progressed from lower to higher forms of life (culminating – that, of course, was the assumption – in mankind). The naturalist Charles Darwin developed the idea of an evolutionary 'tree' in his book *On the Origin of Species* (1859). He suggested that one species could, in time, change into a number of others. These new species would therefore resemble one another (perhaps in the details of their skeletons), and all have links to their common ancestor. An evolutionary tree is a branching diagram that shows how (it is thought) the changes in species occurred. In its most modern form, the tree is constructed using the science of phylogenetics, which deploys the sequences of genes in the organisms to work out how close one species is to another. The resulting diagram is called a phylogenetic tree. It turns out that all species can be plotted on one tree, with a single trunk reaching back to life's origin: the so-called LUCA, or 'last universal common ancestor', of all living creatures. Bacteria, plants, fungi, insects, animals and humans are all descendants of LUCA.

In 1977, while constructing the area of the phylogenetic tree concerning bacteria and similar organisms, two US biologists, Carl Woese (b. 1928) and George E. Fox (b. 1945), considered the position of another group of single-celled micro-organisms called 'archaea', or 'old things' (singular = archaeon). Archaea are similar to bacteria in size and shape, and were formerly considered as a type of bacteria that had adapted and come to occupy environments in which bacteria could not usually survive, such as volcanic springs, where the water is too hot; or lakes, where the salt concentration is too high. Creatures that thrive in extreme environments are called 'extremophiles' (**18**).

The archaea were formerly known as 'archaebacteria': a subgroup of bacteria, in other words. The implication was that archaea might be descendants of bacteria that have evolved to adapt to niche environments. But Woese and Fox suggested that they were fundamentally different from bacteria, with differences in biochemistry, and classified them as a separate 'domain'. What this means is the following.

All life-forms can be classified into groups that become progressively more narrowly defined, ending with a 'species'.

Life in general → Domain → Kingdom → Phylum → Class → Order → Family → Genus → Species.

For example, **Table 3** (overleaf) shows how human beings are classified in a phylogenetic scheme. The groups can be thought of as the trunk, branches, twigs and stalks of the evolutionary or phylogenetic tree; broader groupings are centred on the earlier common ancestors. The definition of the groupings is thus equivalent to a hypothesis of how life evolved on Earth.

The first and most fundamental grouping in a phylogenetic tree is 'domain'. Carl Woese and George Fox proposed that there are three domains: archaea, bacteria and eukarya. 'Eukarya' have cells that have a nucleus. Such cells have a more complicated structure that forms the basis for assembling

Table 3. The Biological Classification of Human Beings

Group	Name	Implication
Domain	Eukarya	Our cells have nuclei.
Kingdom	Animalia	We are animals; for example, we are multicellular, we have stomachs and we move.
Phylum	Chordata	Our bodies have a central axis, in fact, a spinal cord.
Class	Mammalia	We have hair and breasts.
Order	Primates	We have nails, fingers and toes that grasp, and binocular vision: for example, lemurs.
Family	Hominidae	We are great apes, such as Australopithecus and chimpanzees.
Genus	Homo	We have big brains and other human characteristics; for example, Neanderthals.
Species	Sapiens	Like our ancestors Cro-Magnon man, we are modern humans.

complex creatures, for example human beings. But the first two domains are of much simpler life-forms that have a single cell without a nucleus. The key proposal made by Woese and Fox was to separate archaea into a third domain, because they have distinct properties. For example, species of archaea use many more different kinds of energy than eukaryotes. Archaea use chemical compounds, such as sugars, as we do, but also ammonia, metal ions, hydrogen gas, daylight and geochemical energy. They also have some distinctive biochemical components.

Does this identification of a third domain really matter? At first sight, the proposal about the status of archaea may seem rather esoteric, of interest to biologists in the same way that the classification of angels was of interest to a medieval theologian. It is, however, highly relevant to the origin of life.

Because many species of archaea are extremophiles, we can suppose that ancient species of archaea might have been able to exist and develop in the rather more hostile environments that had a larger presence on the early Earth than on today's planet. Archaea can live in a variety of environments, such as salt lakes, the acidic guts inside animals (including us), and volcanic springs, but they thrive especially in the ocean, as a component of plankton, life-forms that drift in the open water of the sea or lakes. They are therefore versatile creatures that could have developed in a great many habitats that were connected as water flowed from one location to another. This versatility and mobility afforded archaea the opportunity to evolve as Earth changed, and to develop into terrestrial life as we now know it. The early archaea evolved into further kinds, and into bacteria, and on to eukaryotes, forming the complex life that this domain includes.

Thus the implication is that present-day archaea are the modern descendants of the earliest life that existed on Earth. We can conclude that, because of their ability to develop in a variety of extreme environments, archaea may well be the life that we find on other new planets.

If we are lucky, we might in the future get to discover worlds similar to the Earth, where there is abundant life of many different forms. On the other hand, we might not find other planets with such rich diversity. We might find a place where life has only just begun, or somewhere hostile, the environment of which challenges the capability of life to survive. We might find a world that started a long time ago, but where life has progressed very slowly. On an extremely cold planet, the development of life-forms may become arrested at a point near its origins, because when organisms are cold, they reproduce slowly (Mars and the planets of the outer Solar System might be such cases, to one degree or another.) Life in any of these places could well consist only of simple archaea.

Science fiction has created the expectation that extraterrestrial life will be humanoid, or at least complex and intelligent. That would be by far the most interesting discovery: life we could converse with. In reality, even if we aim high, we should prepare to fulfil lower expectations.

7

The Spark of Life's Beginning

We know that the chemicals of life are universal, and the fossil record reveals to us what the first living beings actually looked like. But what was the spark that bridged the gap from the inanimate to the animate? How did that first primitive life-form emerge from non-living components? It might be that we have been too conservative in determining what came from space to our planet. Perhaps the seeds of life that were created elsewhere in the remote Universe and transported to other planets, including our own, were more than just the chemicals that have been identified on comets and asteroids. Did living organisms, as well as prebiotic chemicals, arrive from space to seed our planet?

In ancient Greek philosophy, the origin of life was not only a dilemma for the distant past but also a matter for contemporary discussion. According to Aristotle, in his works *The Generation of Animals* and *The History of Animals*, life appears on Earth as a result of spontaneous generation from inanimate matter. He gave several examples, based on reports available to him:

- Spontaneous generation occurred when rain fell on a dried-up pond near Knidos and tiny fishes appeared.
- Eels are not produced by copulation, nor are they oviparous. No eel has ever been caught that had either milt (seminal fluid) or eggs; nor, when cut open, are they found to possess passages for milt or uterine passages. In fact, this whole tribe of blooded animals is produced neither by copulation nor from eggs. That this is so is made absolutely clear by the following: in certain marshy pools, after the water has been completely drawn off and the mud scraped out, eels reappear when there has been a shower of rain.
- Shellfish take shape spontaneously, forming on the side of boats and in many places where nothing of the kind had been present previously.

Although we now know these examples to be false, Aristotle became the authority for natural science in Europe and spontaneous generation was, in medieval times, the widely accepted explanation of the origin of life.

The first serious challenge to this idea was by Francesco Redi (1626–1697), an Italian physician and poet, in 1668. Redi carried out experiments to see if maggots were generated spontaneously in rotting meat, or whether they developed from eggs laid by flies. He put meat in a number of flasks, some open to the air, some sealed completely, and others covered with gauze. Maggots appeared only in the open flasks in which the flies could reach the meat and lay their eggs. At the same time, however, the invention of the microscope revealed that if you put hay in water, teeming numbers of small creatures would 'spontaneously' appear. It was not until the mid-nineteenth century that the French chemist Louis Pasteur (1822–1895) showed that boiling sterilizes meat broth, and for it to become infected it was necessary to expose it to micro-organisms in the air. Pasteur summarized his experiments in the words *La génération spontanée est une chimère*: 'Spontaneous generation is a myth'.

These experiments show that life usually propagates from other forms of life. Seeking to explain how it may originate from inanimate chemicals, therefore, is a high hurdle to leap. Charles Darwin himself found the question of how life came into existence from inanimate matter to be very difficult. He wrote in 1863 to his friend, botanist J. D. Hooker (1817–1911), director of the Botanical Gardens at Kew, about an unfavourable summary of his views that had just been published:

> It will be some time before we see 'slime, snot or protoplasm'
> generating a new animal. But I have long regretted that I truckled to
> public opinion and used [the] Pentateuchal term of 'creation', by
> which I really meant 'appeared' by some wholly unknown process.–
> It is mere rubbish thinking, at present, of [the] origin of life; one
> might as well think of [the] origin of matter.

Darwin's words reveal that he had tried to provide a tone to his account of the origin of life that would make it acceptable to the religious, by using terminology reminiscent of the description of the same process given in the Book of Genesis. He did not, at this time, realize that his underlying idea would be so strongly resisted. He was right that the origin both of life and of matter would remain problematic for a long time. More than a century later, we still do not have an entirely coherent explanation.

The problem of life's beginning persists in our present-day picture of the origin of the planet. The Earth was formed from the accumulation of material in the solar nebula, with small embryonic planets, or 'planetesimals', crashing together and building the Earth into the size it is today. The collisions released massive amounts of heat energy, and it is estimated that at its birth the surface temperature of the Earth was approximately 2,500°C. Given this heat, the Earth's surface was certainly sterile at that time. So how did life arise here?

* * *

One attractive idea, which postpones explaining the true origin of life on Earth, is that just as eggs are introduced into rotting meat by such external agents as flies, life was introduced to Earth from outer space. The 'panspermia' hypothesis, which was developed in the early years of the twentieth century by the Swedish chemist Svante Arrhenius (1859–1927), suggests that life is widespread throughout the Universe, and that it is transported from one place to another. For example, dormant bacteria or archaea could be propelled through space by the pressure of stellar radiation, or travel on meteoroids and comets that have been ejected from a place where life already exists. If a meteoroid falls to a planet such as Earth, life may seed on the planet's sterile surface, developing and evolving if it finds a suitable environment.

Astronomers know that, in our Solar System, meteorites do bring rocks from the surface of one planet to another. Three other worlds have been identified as sources of some of the meteorites that have fallen to Earth.

About thirty-four meteorite falls have been identified in which the rock had its origin on Mars, shot into space by the impact of an asteroid on the surface of the Red Planet. They are collectively known as SNC (pronounced 'snick') meteorites, after the initials of the names of the first three examples that were identified. Meteorites are conventionally named after the post office located nearest to their fall. The first known of the thirty-four SNC meteorites was the Chassigny, which came to Earth with sonic booms on 3 October 1815 near Chassigny, in the Burgundy region of France. A man working in a nearby vineyard saw something fall from the sky with a hissing sound, leaving a smoking trail, and ran to see what it was. In a small hole in the freshly ploughed ground, he collected stones, hot to the touch, which proved to be meteorites.

The Shergotty meteorite fell to Earth in similar circumstances on 25 August 1865 in Shergotty, in the state of Bihar, India. This type of meteorite is the most common of the Martian meteorites, made of a distinctive mineral called shergottite.

The third Martian meteorite fall consisted of forty pieces that landed close to Nakhla, near Alexandria, Egypt, on 28 June 1911. It was reported to have produced a white column-like cloud and explosions that frightened local residents, one of whom, a farmer named Mohammed Ali Effendi Hakim, is said to have seen a piece strike and kill a dog, 'leaving it like ashes'. This dramatic, graphic and much-repeated account, which would describe the only non-fictional example of an 'earthling' killed by a Martian, is, sadly, the exaggerated product of a lively imagination, on the evidence of another contemporary report by a British diplomat who investigated the incident. But the 10 kg of black fragments found in the fields of the farmland around the village among the okra, cucumbers and strawberries are real. They are prosaic in their appearance, but the history of these rocks, as revealed by science, is mind-boggling.

The distinctive journey of the SNC meteorites gradually became clear when it was understood that they were different from the thousands of other meteorites that had been collected. As measured by the rubidium–strontium dating method (**Table 2**, p. 73), nearly all of the SNC rocks have a common crystallization age. They were last molten and solidified 1,370 million years ago – much more recently than most meteorites, which solidified 4,000 million years ago or more. Some of the minerals in the SNC rocks had weathered in the presence of water, not in a dry desert. The chemical composition of the SNC meteorites is similar to analyses of rocks by spacecraft on the surface of Mars, made in 1976 by the Viking landers. This hinted that the meteorites came from Mars. The decisive fact came from a shergottite, picked up in 1979 in Antarctica.[1] Numbered EET79001, it was, in 1983, found to contain bubbles of gases trapped within its glass-like material, the composition of which closely resembled the atmosphere of Mars (again according to analysis by the Viking landers). This established in everyone's mind that the SNC meteorites came from Mars, from a magma field that solidified after a volcanic eruption 1,400 million years ago.

The most comprehensive explanation is that most of the SNC meteorites probably launched into space after a big piece was ejected from Mars in an impact that happened 200 million years ago. The big piece was broken into smaller pieces by a collision 10 million years ago, while it was orbiting in the Solar System. The small bits were then exposed to 'weathering' in space by cosmic rays; the cumulative effect of these rays on the rock material is what tells us that the individual pieces of SNC meteorites were in space for 10 million years.

It is estimated that the original impact made a crater on Mars about 100 km in diameter, but this precise crater has yet to be definitively located. There are some elliptical craters on Mars, formed from a grazing, oblique impact, and it is easy to imagine that from this, some material might have been hurled forwards into space. One such crater, called Rahe, is located on the slopes of the giant Martian volcano known as Ceraunius Tholus (**19**). Rahe has been suggested as the source of the SNC meteorites. The clinching observation would be to identify a volcano, the magma from which is of the same composition and age as the SNCs that have been found. This would best be done by analysing samples of Martian rock brought back to Earth by geologist-astronauts. Until this becomes possible, we have to be content with remote sensing; there is some encouraging evidence from the Mars Global Surveyor and Mars Odyssey satellites, which have identified areas with the right sort of minerals. The Nili Fossae region of the Syrtis Major volcano and the crater Zumba near the Tharsis volcano are promising possibilities.

There are two other worlds that have supplied individual meteorites to Earth by a process of interplanetary transportation. Approximately 160 meteorites spread across about 25 falls have landed on Earth from the Moon. It is much easier (and less expensive!) to pick these lunar rocks up from the ground, here, than from the Moon's surface, even if some of them fell in Antarctica, which is not the simplest place to which to mount a scientific expedition. But of course the reason that we are able to identify the lunar origin of these meteorites is that their composition matches that of rocks brought from the Moon back to Earth by the Apollo astronauts.

A large number of meteorites – about 5 per cent of all falls – originated from the asteroid Vesta, although the evidence for this source is not quite as secure as it is for those that came from Mars and the Moon. Scientists have examples of material from these two places in their laboratories with which to compare the composition of meteorite material. The composition of Vesta is known only from the colour of the light that it reflects. The distribution of light is called a spectrum, and the spectrum of a particular class of meteorite, called HED meteorites, matches that of Vesta. (HED stands for 'Howardite-Eucrite-Diogenite', three related minerals that are found in these meteorites.) Astronomers think that HED meteorites were probably ejected into space from Vesta in the meteor impact that created a large crater on the asteroid. Located at its south pole, the crater has been imaged by the Hubble Space Telescope and, in 2011, by the Dawn space probe (**12**). The crater is 30 km deep and 460 km across. Considering that Vesta is only 530 km across, this impact came close to breaking it up completely, and certainly ejected a large amount of debris into space. Some of this debris still orbits in the Solar System as small asteroids, known as the Vesta family.

It does not seem very likely that there is life on Vesta when none has been found on the Moon, but if life exists or once existed on Mars it could have been transported to Earth via meteorite falls. Of course, life could also be transported in the opposite direction, from Earth to Mars (or the other planets), by the same natural process.

Life could also be transported from our planet to others by means of space probes, launched from Earth to explore other worlds. This is an ethical issue, as introducing foreign biology to another world could compete deleteriously with the native biosphere. Unfortunately it is true that we have already contaminated one such world; human waste was left behind when astronauts travelled to the Moon, not to mention machinery and other artefacts that had had the chance to become contaminated on Earth or by the astronauts. It is also a practical issue, in that if we accidentally introduce terrestrial life to a planet, and subsequently find evidence of living organisms there, it will be impossible to determine where the life came from – did it develop there or did it originate on Earth? For this reason, there are rigorous protocols for the treatment of spacecraft that are destined to land on other planets, and space agencies have implemented sterilization practices for planetary protection.

Again, by symmetry, any space mission that returns with material from another planet might carry extraterrestrial organisms. For this reason, space missions that do this are subjected to stringent isolation procedures, to protect Earth from invasion by extraterrestrial life. A NASA plan for a future space mission that will return to Earth from Mars with Martian samples is that it will be held at the International Space Station and quarantined before being carried carefully to Earth. This could be regarded as an overcautious gesture to allay public concern, because bits of Mars have – throughout the entire life of the Solar System – already travelled to Earth in an uncontrolled way, and the biospheres of the two planets overlap as components of the larger biosphere that is the Solar System.

* * *

It is a fact that rocks can be transported across interplanetary space from one planet to another; it is conjecture that they can travel between planetary systems. What we can say is that many asteroids (and comets) are loosely held in the Kuiper Belt and Oort Cloud, in the far reaches of our own Solar System. They are tied only weakly, by gravity, to the Sun. Some of these bodies must have escaped into space, after being given a kick by a close encounter with another one. Estimates suggest that by far the majority of the comets and asteroids were ejected into interstellar space before they ever got to the Oort Cloud and Kuiper Belt, when the planets flung them out of the Solar System.

Eventually, nearly all the comets will leave our Solar System and drift off into interstellar space. Throughout its life as a star, the Sun has been losing mass, and in the future it will do so more rapidly than now. Its ties on the comets will get looser, and they will all eventually detach from the Solar System.

If comets leave our Solar System to enter interstellar space, then conversely some must travel the other way from time to time, perhaps even impacting onto the Earth. Such an interstellar comet would approach the Solar System very quickly, falling in on a hyperbolic orbit, as distinct from the more usual elliptical or parabolic one.[2] A number of comets have been observed with orbits that do indeed appear hyperbolic, but only just. It could be that a combination of errors in measurement and some small effect, such as the nudge caused by the jets of material that spurt from the comet, could account for these. Such comets are speeding, but like a crafty motorist, they are only just above the speed limit.

If interstellar comets were to reach Earth, their composition would be typical of the planetary system where they originated, rather than of our Solar System. Attempts to identify 'foreign' comets over the past few decades have all been negative, or at best equivocal.

* * *

The upshot is that we are unsure whether or not comets enter the Solar System from interstellar space; therefore we cannot know for certain if in the past any might have brought life to Earth. There is, however, one further tantalizing hint that suggests this might be true. As explained in chapter 4, amino acid molecules are chiral, and life on Earth uses left-handed forms almost exclusively. There does not seem to be a fundamental reason why this is so; it is possible to imagine life-forms built with the right-handed forms of amino acids, although not with a mixture.

So where does the handedness of terrestrial life come from? If a racemic mixture of amino acids is exposed to particular kinds of polarized light, the molecules of one handedness tend to switch to the other. Such light is radiated by pulsars: neutron stars in space produced by supernova explosions. Suppose, therefore, that a comet covered with a racemic mixture of amino acids swept past a pulsar. The amino acids would convert to a particular handedness. If the comet delivered this interstellar cargo to the Earth, life could begin from this point and continue to evolve using this same chirality. This remarkable speculation about the origin of life would suggest that it was the chance result of an interstellar comet's trajectory.

To some extent the panspermia hypothesis merely removes the problem from the Earth and transfers it somewhere else, but of course if the Universe has been eternal, the question of how, where and when the first life began is pushed back indefinitely. The hypothesis survives in a theory first presented in the 1980s, and subsequently propounded by British astronomers Fred Hoyle (1915-2001) and Chandra Wickramasinghe (b. 1939). They identified various features in the infrared spectrum of the materials in interstellar space with similar features in the spectrum of viruses and bacteria, implying that the same materials are present in both. These two scientists attribute waves of epidemics among humans on Earth not to mutated versions of viruses and bacteria of terrestrial origin, but to showers of previously unknown viruses and bacteria from space. This in itself is an idea with a long history, going back to the attribution of widespread illness to the influence of comets and other astral phenomena, a branch of astrology particularly prevalent in medieval Europe between 1450 and 1700. We have a linguistic fossil left from this time. The word 'influenza' was imported into English to refer to an epidemic that swept Italy in 1743 at the time of the appearance of a bright comet. Few astronomers, and fewer biologists, give the Hoyle-Wickramasinghe theory much credence.

* * *

An alternative to panspermia is the idea that organic chemicals and the life that formed from them developed on Earth in some purely chemical process. Presumably, similarly to making a cake, one needs ingredients, a recipe, and an oven fuelled by a source of energy. One early idea, sketched in 1908-11 by the father-and-son team Thomas C. Chamberlin (1843-1928) and Rollin T. Chamberlin (1881-1948), suggested that 'planetesimals' – the small-sized bodies that merged into the planets – were the source of organic material from which life evolved. 'The planetesimals are assumed to have contained carbon, sulphur, phosphorus and all the other elements found in organic matter,' they wrote, 'and as they impinged more or less violently upon the surface formed of previous accessions of similar matter, there should have been generated various compounds of these elements.' The Chamberlins had offered suggestions about the ingredients and the oven, but did not give much detail about the recipe.

Darwin himself proposed a solution about the origin of life in another letter to J. D. Hooker, in 1871:

> It is often said that all the conditions for the first production of a
> living organism are now present, which could ever have been present.

But if (and oh! what a big if!) we could conceive in some warm little pond, with all sorts of ammonia and phosphoric salts, light, heat, electricity, etc., present, that a protein compound was chemically formed ready to undergo still more complex changes, at the present day such matter would be instantly devoured or absorbed, which would not have been the case before living creatures were formed.

The idea that energy input into a soup of primitive chemicals commonly found on Earth in its early stages could create protein compounds 'ready to undergo more complex changes' was quickly discovered to be problematic. The issue was that oxygen in the Earth's atmosphere stops the development of complex organic chemicals that contain atoms of carbon and hydrogen, by readily combining with them to make carbon dioxide and water. Oxygen is a very reactive gas and holds strongly to carbon and hydrogen, so once it has combined with these elements, it is hard to go back.

The British-born polymath J. B. S. Haldane (1892–1964) and the Soviet astrobiologist Alexander Oparin (1894–1980) provided the answer – in principle – in the early twentieth century. When Earth began, its atmosphere must have been oxygen-free. The principal constituent of the Universe is hydrogen, and it is easy to understand that this element might once have formed the main constituent of the Earth's atmosphere. But, if indeed this was ever so, the hydrogen is soon lost – it is the lightest element and its atoms move quickly, readily flying off into space. Carbon and nitrogen are two very abundant elements in the Universe, and hydrogen combines with carbon to make quite heavy gases. These gases and nitrogen could well have composed the earliest manifestation of the Earth's atmosphere. The geochemist Harold Urey (1893–1981) from the University of Chicago specifically proposed in 1953 that the atmosphere of the Earth would, at its origin, have consisted of water vapour, ammonia and methane – with no presence of oxygen.

In lectures on the subject at the university, Urey proposed an experiment to see what would happen should lightning spark in such an atmosphere. His suggestion was taken up by a biochemistry student, Stanley Miller (1930–2007). Miller approached Urey with his request, but as he later recalled, 'The first thing he tried to do was talk me out of it.' Urey was discouraging; he expected that it would take much longer than the three years typically available to a PhD student to do anything of consequence. But Miller persisted, and set up a flask of water, to represent the oceans, and connected it to a flask of gases through which he passed electrical discharges to simulate lightning. He set up the varying temperature of the water in such a way as to cause evaporation and condensation, simulating the water cycle: from sea to clouds to rain and back again. After just two days, glycine, a simple amino acid, emerged, and by the end of the week about a dozen more amino acids had been created. The water took on a brown tinge, formed by other tarry chemical compounds.

Amino acids are organic molecules containing (a) carbon atoms (the property that makes them 'organic'); (b) an amine group (that is, the molecule contains a segment that has three groups of atoms centred on a nitrogen atom); (c) a carboxylic acid group (that is, the molecule contains the –COOH group of atoms: two atoms of oxygen flanking one of carbon, an atom of hydrogen attached to one of the oxygens, and the whole group attached at the carbon atom to the rest of the molecule); and (d) a further chain of atoms that defines the type of amino acid. A general formula for a simple amino acid is $H_2NCHRCOOH$, where R is a general organic chemistry group. Amino acids are critical to life. They link into chains, as proteins. Twenty kinds of amino acids are encoded for on the DNA molecule, the genetic basis for all kinds of life. Amino acids are assembled into proteins that build the architecture of a given organism. Out of inanimate gases, Miller's experiment made organic molecules that could truly be said to be the basis of life.

The reaction to Miller's work in the science community was, at first, one of profound scepticism. Miller recalled:

> One scientist was sure that there was some bacterial contamination of the discharge apparatus. When you see the organic compounds dripping off the electrodes, there is really little room for doubt. But we filled the tank with gas, sealed it, put it in an autoclave[3] for 18 hours at 15 psi [to purge the tank of extraneous compounds]. Usually you would use 15 minutes. Of course, the results were the same.

Since Miller's original work, the experiment has been repeated hundreds of times with different gas mixtures and various forms of energy that simulate the sources available on the primitive Earth, such as ultraviolet light and volcanic heat. The results are always the same: biochemical compounds are produced, provided the gas mixture representing the atmosphere contains no oxygen. Miller's original, simple experiment is one of the most – if not the most – successful in creating life-giving chemicals. 'The fact that the experiment...works and is so simple is what is so great about it. If you have to use very special conditions with a very complicated apparatus there is a question of whether it can be a geological process,' said Miller.

Miller's experiment envisaged that life could have started on Earth from simple ingredients subjected to rather energetic cooking by lightning. Others have postulated that the ingredients might have been partly cooked into organic chemicals in space, delivered to Earth and then reheated for the final process in which life was made.

How did organic chemicals get made in space? Life is based on carbon, an element that is made in stars and is spread when the stars explode as supernovae (20) everywhere in the Universe. Radio astronomers have discovered about 150 varieties of organic molecules in interstellar space. A common manufacturing scenario is that they are made on the surface of interstellar grains floating in space. What happens is that individual atoms in space encounter a grain and stick to its surface. If other atoms land close by and also adhere to the grain, the atoms can combine and make a molecule, which then floats back off into space. The molecules that have been detected include simple gases of only a few atoms, such as carbon monoxide (CO) and methane (CH_4), but also more complicated ones, for example methanol (CH_3OH) and acetone ((CH_3)$_2CO$). There are reports of the detection of glycine (H_2NH_2CCOOH), the first amino acid synthesized by Miller's experiment.

Glycine and other amino acids have been found in meteorites. The origin of these chemicals is not clear. Perhaps the meteorite assimilated them by absorbing them from the general material raining into the Solar System from interstellar space; or perhaps the chemicals were made on the grains of the solar nebula, which congealed into the meteorite as it was formed in interplanetary space. The Murchison meteorite (chapter 4) contained scores of amino acids, including glycine. Comets are covered with a tarry crust that contains many organic compounds. In 2004, the space mission Stardust passed through dense gas and dust surrounding the icy nucleus of the comet Wild 2.[4] As the spacecraft flew through this material, a special collection-grid filled with aerogel – a novel sponge-like material that is more than 99 per cent empty space – gently captured samples of the comet's gas and dust. The grid was stowed in a capsule that detached from the spacecraft and parachuted to Earth in 2006. In 2009, a team of NASA scientists, led by Jamie Elsila (b. 1974), revealed that they had found cosmic glycine in the aerogel and on the collection-grid. It was clear from a close analysis of the individual atoms in the glycine that this substance was not a terrestrial contaminant: it had been made in space.

* * *

Darwin used the words 'warm pond' to describe the possible place in which a chemical mixture is simmered to develop more complex molecules that transition into life-forms that might be something akin to present-day archaea. We could regard the pond as a metaphor for whatever the actual location was – indeed there might have been many such places that provided starting points for life to trickle from, later to merge into a single evolutionary river.

Deep-sea hydrothermal vents are a possible location for the origin of life. A hydrothermal vent is a fissure in the rock that ejects water heated from geothermal sources below ground. Examples include springs, fumaroles and geysers, such as the ones found in Yellowstone National Park, Wyoming. 'Black smokers' are hydrothermal vents under the sea; the first examples were discovered by the *Alvin* undersea exploration vehicle in 1977, in the east Pacific Ocean. Black smokers are the centres of a flourishing community of life at the bottom of the sea. The water that comes out of these vents is rich in dissolved min-erals, including sulphur compounds, which are used as food by bacteria that use volcanic energy to process it. The bacteria sit at the bottom of a complex, densely populated food chain. These habitats are different from those at the surface of the Earth, where the primary input of energy is sunlight. No sunlight penetrates to the depths at which the black smokers are found, as far down as 5 km below sea level. Some of the organisms in the habitat around black smokers depend upon oxygen, produced at the surface by vegetation or cyanobacteria using photosynthesis, but others do not use oxygen at all.

According to this theory, life on Earth began deep on the floor of the ocean, springing from an extreme environment of hot, hellish, brimstone-smelling, dark volcanic vents, before migrating and evolving into the more benign circum-stances and fresh light of the surface world. Life as we experience it was not due to a fall from a Garden of Eden into the painful tribulations of our present world. It was a rise from a noxious hell into an earthly approximation of paradise.

In other theories, scientists have suggested that life might have developed along the shorelines of oceans, where the chemical soup could have been concentrated in shallow lagoons. The lattice structures of clay minerals on the damp lagoon shores may have been vital as templates for the organization of simple organic molecules into larger ones. Strong tides caused by the Moon (which at that time was much closer to Earth) may have played a role in concentrating the soup. The tides might also have concentrated grains of uranium, providing radioactive energy that could have been the 'spark of life'. Or perhaps the spark was strong ultraviolet light from the young Sun, readily penetrating the Earth's atmosphere because there was not yet an ozone layer (since the free oxygen that forms the ozone is only released through living organisms photosynthesizing). Washed back out to sea, these materials might have sunk to the sea floor, found a favourable environment in the region of the black smokers, and developed into such primitive living creatures as archaea and bacteria.

How this last crucial step took place is the most obscure part of the process. Did the complex biochemicals created by chemical processes develop into even more complex biochemicals, which included nucleic acids and might have been simple forms of DNA that have the ability to reproduce and then afterwards develop into living cells? Did the complex biochemicals organize themselves into differentiated cell-like structures and then develop ways in which they could reproduce?

In 1951 the physicist and biologist John Bernal (1901–1971) coined the term 'biopoesis' for the process by which living matter might evolve from inanimate matter, and listed its three stages:

- The origin of simple organic chemical molecules.
- The origin of complex organic chemical molecules of the sort used by living organisms.
- The evolution from chemical molecules to cells.

The first stage seems relatively clear; the Universe is full of simple organic molecules made in interstellar space, delivered to Earth in a variety of ways. The clue to the second stage is in Miller's experiment and its repetitions, but which exactly of the several possible chemical mixtures, locations and energy sources was actually responsible for the origin of life on Earth is still to be discovered, as is whether there was one or many such events. The materialistic opinion of most scientists is as expressed by the physiologist Edward Schäfer (1850–1935), president of the British Association for the Advancement of Science, in his presidential address in Dundee in 1912, which caused a sensation in the newspapers. '[T]he problems of life are essentially problems of matter; we cannot conceive of life in the scientific sense as existing apart from matter.... [L]iving matter must have owed its origin to causes similar in character to those that have been instrumental in producing all other forms of matter in the universe.' That essential third stage, however – the recipe by which life is started – remains mysterious. Darwin's 'warm little pond' may have been an understated metaphor for the steam-heat of black smokers on the deep ocean bed, but precisely how this environment sparked the existence of life is still unknown.

8

Evolution of a Planet

How likely is it for conditions to occur on a planet that enable life to develop? The case of Earth suggests that a life-friendly planet is an exceptionally rare thing. Could it even be unique in the Universe? Or do planets and life evolve together in a mutually beneficial way?

The biologist James Lovelock (b. 1919) put forward in the late 1960s the concept that the Earth and its life together constitute a single self-regulating entity, which has maintained itself since life developed here, keeping Earth's climate and chemical composition comfortable for organisms. This is the Gaia hypothesis, named at the suggestion of Lovelock's friend, the novelist William Golding (1911–1993), after the Greek goddess of the Earth, Gaia. The theory has a superficially New Age flavour, which initially resulted in a strong negative reaction from other scientists, but, as its features became better understood and some of its predictions proved to be accurate, especially those regarding how the climate is regulated, the theory has grown to be more acceptable. In particular, in 2001, in what became known as the Amsterdam Declaration, the scientific communities of four international global-change research programmes[1] used the Gaia hypothesis as the basis for their sober call for action to tackle climate change caused by human activities. The Declaration described the science behind the environmental concerns in the following way:

> The Earth System behaves as a single, self-regulating system
> comprised of physical, chemical, biological and human components.
> The interactions and feedbacks between the component parts are
> complex and exhibit multi-scale temporal and spatial variability.
> The understanding of the natural dynamics of the Earth System has
> advanced greatly in recent years and provides a sound basis for
> evaluating the effects and consequences of human-driven change....

Global change cannot be understood in terms of a simple cause-effect paradigm.... Changes cause multiple effects that cascade through the Earth System in complex ways. These effects interact with each other and with local- and regional-scale changes in multidimensional patterns that are difficult to understand and even more difficult to predict.... The Earth System has operated in different states over the last half-million years, with abrupt transitions (a decade or less) sometimes occurring between them.

The Gaia hypothesis highlights the way that a planet and the abundant life that it may host can change and develop together. It is an optimistic view: if the environment on a planet changes towards some deeply hostile state, life will respond, evolve and modify the environment back to something that is more benign. Lovelock suggests that this will happen as a result of manmade global warming. We humans may suffer, but – literally – life will go on.

An alternative view is more negative. Life may create the conditions for its own extinction, proliferating and modifying its environment in such a way that life in general becomes impossible. The English thinker, the Reverend Thomas Robert Malthus (1766–1834), was challenged by his father to explain why he thought that a world of benign peace was impossible in the long run. He wrote a paper, published and republished between 1798 and 1826, titled 'An Essay on the Principle of Population; or, a view of its past and present effects on human happiness; with an enquiry into our prospects respecting the future removal or mitigation of the evils which it occasions'. In an increasingly industrialized England, at a time when the population was expanding spectacularly, Malthus recognized that the population grew in proportion to the number of people who were alive to have children – a mathematical form known as an exponential increase, which rapidly runs away – but that the food supply grew at a much slower rate. He concluded that sooner or later the population would collapse, because of poverty, famine and consequent disease. This pessimistic view gave rise to the adjective 'Malthusian'.

Malthus's views were very influential on Charles Darwin, and are still regarded as a valuable simplification of the processes that give rise to the oscillation of the population of a species within a closed environment with a limited food supply. To the extent that human beings are consuming roughly three-quarters of the available resources of our planet, we should be worried about Malthusian ideas. We have been saved from the consequences of the mathematics so far, not because the mathematics is wrong, but by the ingenuity of humankind in limiting our own population, developing the use of existing resources and finding new ones. These significant achievements have perhaps only postponed the outcome of the maths.

* * *

These differing views on the interaction of life and its host planet emphasize the effect that life has on the evolution of a planet, and vice versa. We can map the scope of the interaction by studying the development of life on the Earth and the changes that have occurred.

For the first 1,000 million years of our planet's existence there was very little life on Earth. This period is termed the Hadean eon, named by the US geologist Preston Cloud (1912–1991) in 1972, after the desolation of Hades, the Greek underworld where there is no life. The Hadean eon itself may have been lifeless, but what happened some time towards the end of this period had profound effects on the subsequent development of life on Earth.

The Hadean eon began with the formation of the Earth. The condensation of the Earth in space in the solar nebula happened over perhaps 10 million years. Dust grains in the solar nebula collided and stuck together, coagulating into bodies up to about 1 km in diameter. At this size, the lumps, or planetesimals, were large enough to attract smaller lumps in their vicinity, and of course the larger a planetesimal grew, the more material it was able to pull on to itself and thus expand further. By the time that such a planetesimal had grown to about 1,000 km in size, it was able to sweep up more dust and smaller planetesimals in its passage through the remains of the solar nebula. It may have grown by gathering material, in the manner of a sweeping broom, through physical contact. Alternatively or additionally, it may have grown through a more subtle accretion process in which dust flowed in orbit either side of the intruding planetesimal, the two streams of dust collided in space at the rear of the planetesimal, and the dust fell onto its back surface.

In the cool, far reaches of the outer Solar System, beyond the 'snowline', there formed the four giant planets – Jupiter, Saturn, Uranus and Neptune – and their larger satellites. The accretion there was to a great extent indiscriminate, and the giants collected everything. In some ways these planets are like stars – for example, they are predominantly gaseous and support themselves by means of an upwards internal pressure that counteracts and balances the downward force of gravity – but because they are smaller than stars, and therefore cooler and less dense, the normal nuclear fusion processes that take place in stars do not occur, so they remain planets.

The giant planets were, at that time, more closely spaced than now, but none the less more widely separated than the planets that formed in the inner

Solar System. Pieces of planetesimals orbited thickly in the outer zone, which stretched out well beyond the furthest giant planet, Neptune. The pieces were numerous in the outermost parts of the Solar System; everything moves much more slowly in the far reaches of the Sun's influence, and the build-up of planets took a correspondingly long time, with an abundance of material left over.

In the warmer, inner zone of the Solar System, the four terrestrial planets formed from solid, dusty material left within the 'snowline'. The violent fall of material onto each planetesimal released energy that heated its entire mass, melting the iron, nickel and other metals that it contained, the liquid metal draining down into the planetesimal's core. The lighter material was left as a cooling slag-like mantle that built up on the planetesimal's surface.

It is likely that all planetary systems begin with the same division into an inner and an outer half, as in our own Solar System. What happens next is much more chaotic, and for our Solar System, this was the stage that gave it distinctive properties that may not be replicated again, elsewhere.

The inner Solar System became dominated by just a few hundred embryonic planets. The orbits of the small planetesimals had been circular as they rotated in the solar nebula, but the mutual interactions of the large and numerous embryonic planets during their final stages made these orbits become chaotic, and increasingly eccentric, causing the paths of adjacent planets to intersect. Collisions were inevitable, during which planets merged. A relatively small number of planets resulted, perhaps five or six, relatively well spaced, each having largely swept clean its adjacent feeding zone. Now that there were so few planets, each settled into an only slightly eccentric orbit, and collisions between the inner planets became rare to non-existent.

One of the last major collisions was between the embryonic Earth and a Mars-sized large planetesimal, or small planet, dubbed Theia.[2] Theia struck the proto-Earth a glancing blow, scattering much of the mantle of each body into space, while the two liquid metal cores coalesced into one planet, like raindrops running together on a windowpane. The glancing blow caused the larger body, the Earth, to rotate more quickly, while the mantle material flew into orbit around the Earth, condensing over time into a coreless second planet: the Moon. This hypothesis about the origin of the Moon is informally called the Big Splash.

The extra-large iron core that Earth inherited during this collision, and its high rotation speed, had profound effects on the evolution of life. Because the core is so substantial, it has remained warm and liquid as a result of energy released by the decay of abundant radioactive elements within. The warm liquid rises and falls in strong currents through convection, which causes the Earth to have a strong magnetic field. The core interacts with the more solid mantle above, and the lower depths of the mantle become much more malleable, like plastic. This facilitates the movement of the solid material above and makes possible plate tectonics, or continental drift, with significant effects for the development of life here.

*   *   *

Continental drift is the movement of the continents over the Earth's surface. The notion was suggested on the basis that the coastlines of the individual continents fit together like a jigsaw puzzle, and therefore that they must at some point have been ripped apart. The idea was developed by the German geophysicist Alfred Wegener (1880–1930) in 1912, but met initially with a hostile reception. A range of multidisciplinary but circumstantial evidence was built up, however, including compatible rock types, magnetic structures and fossil content on either side of the boundaries of the continents, where it looked as if they had fitted together. In the 1960s, the mid-Atlantic rift was discovered, with clear evidence from the magnetic properties of the sea floor either side of the rift to suggest that the floor was spreading outwards from the centre, and that therefore the American and European/African continents were separating. This was all integrated into the theory of plate tectonics, which suggests that the rocky surface of the Earth is made up of a number of 'plates' that move independently. They separate from, collide with, or rub against one another. As explained above, the plates float on the more plastic interior zones of the Earth, and are driven in their motions by a combination of forces, including the convection of the Earth's liquid-iron core. The movement of the crust is facilitated and lubricated by water; if there were not such an abundance of water on the Earth, its tectonic plates would experience more friction and continental drift could be arrested.

Tectonic plates form and re-form as material rises up from and is pushed down into the lower parts of the Earth's crust. Throughout Earth's history they have repeatedly aggregated into large 'supercontinents' and then fragmented for a period in which they slipped about the Earth's surface, eventually colliding and sticking together again. The typical timescale for this cycle is about 500 million years, so there have been several supercontinents in the history of the Earth (**Table 4**, overleaf). The earlier ones are more difficult to separate or identify, since their pieces ('cratons') have been mashed up or incorporated into subsequent supercontinents, like old pieces of building material used and reused in a series of mosaic floors.

Table 4. Supercontinents

Supercontinent	Age (millions of years ago)
Vaalbara	3,500–3,000
&/or Ur	3,000
Kenorland	3,000– 2,000
Columbia	2,000–1,500
Rodinia	1,000–750
Pannotia	600–500
Pangaea	300–150

Note to Table 4. Although the broad concept of supercontinents is sound, and geologists generally agree on the way that the most recent ones have been assembled and then broken up, the evidence about the earlier supercontinents is literally fragmentary. Vaalbara and Ur are two of the more hypothetical supercontinents, which may have existed one after the other or may be alternative interpretations of the evidence.

It was on and in the oceans around the last two supercontinents and, later, on their fragments that complex life developed. The most recent supercontinent was Pangaea, which broke into two pieces, called Laurasia and Gondwana, about 175 million years ago. About 150 million years ago, Gondwana separated into Africa, South America, India, Antarctica, and Australia. Then Laurasia split North America from Greenland and Eurasia. The separation of Pangaea into continents allowed life to evolve on each of them in similar niches but in different ways, even though the species may share a common ancestor. For example, marsupials evolved on South America and hopped to Australia via Antarctica, while placental mammals on North America and Eurasia outfought and displaced the marsupial populations. The fragmentation of the supercontinents is one of the reasons for the diversity of life on Earth.

Tectonic motion on the scale it happened and still happens on Earth is not common on other planets, and, as far as we can tell, continents do not drift on most of them. Life on other planets like Earth will therefore have much more uniformity. Because of the variety of species on Earth, it is logical for us to have handbooks of the life in Africa and in Australia, for example, even though the two areas have very much the same ranges of habitats. A similar geographical division might not make as much sense on another planet.

The fragmented continents drift on the Earth and presently concentrate in the northern hemisphere, but form strong north–south barriers to the flow of the oceans around the world. Regional climates are heavily influenced by the oceanic currents: for example, the climates of the west of Europe and the north-west of America are strongly enhanced by the clockwise gyres[3] of warm currents in the Atlantic and Pacific Oceans, the Gulf Stream and the Kuroshio Current, respectively.

Of course, not everything produced by plate tectonics is benign. Although water lubricates continental drift, tectonic plates do not slip past one another smoothly. They stick, and build up tension that is then suddenly released. This is the reason for earthquakes, which can cause immense local destruction of habitat through sudden motion, as well as landslides and tsunamis. Human mortality through earthquakes is considerable and, given the general increase over time in the human population in large cities on seashores that are vulnerable to both earthquake shocks and tsunamis, the first earthquake to cause a million deaths will probably occur in the not-too-distant future.

* * *

The second effect of the large iron core created by the collision of Theia with the Earth is the strength and relative longevity of the Earth's magnetic field. This magnetic field is caused by the convection of the liquid-iron core: the combination of the convective motions with the rotation of the Earth creates a magnetic field through a dynamo effect. The magnetic field defends the Earth from the impact of charged particles from the Sun (chapter 10). This planetary defence shield pushes back charged radiation, preserves our atmosphere and reduces the cosmic radiation level at the surface of the Earth, stopping solar particles from scouring away the atmosphere and thus keeping the surface benign for living organisms. On a planet without a strong, long-lasting magnetic field, life will be driven into caves and fissures that shield it from harm, or life might even be impossible. We can see one such probable outcome in the planet Mars (chapter 13).

It is astonishing to think that the habitability of some regions of the planet, the devastation of earthquakes and the fact that we still have air to breathe can all be traced back to an interplanetary collision 4,000 million years ago.

* * *

What happened in the Solar System after the collision of Theia and the Earth is not certain, but an interesting scenario has emerged in the years since 2005, as a result of what is known by astronomers as the Nice Simulation, by Rodney Gomes, Hal Levison, Alessandro Morbidelli and Kleomenis Tsiganis, an international group of mathematicians centred on the Côte d'Azur Observatory, Nice. According to the Nice Simulation, what happened in the next 500 million years or so of the history of the Solar System was a gigantic game of interplanetary billiards.

Exactly what happened is uncertain for the following profound reasons. The Nice Simulation is an accurate calculation of how the planets moved and interacted with each other during that time. But, although computers are now so advanced that the calculations are accurate, and it is believed that astronomers know the theory of gravity well enough for their formulations in the computer to be realistic, the motions of the planets did not happen exactly in the way that is calculated. The calculations as a whole do not give the right answers when astronomers attempt to predict where the planets were a long time ago in the past (or where they will be a long time in the future). The reason is that the orbits of the planets are 'chaotic'. If you displace the starting position of one of the planets by just one centimetre, you might expect the planet to be about the same amount different from its calculated final position 10 to 100 million years later. If so, the change of outcome would be immaterial. But in fact, the way that gravity works and the very nature of the equations mean that the planet could be literally anywhere else in its orbit. That one centimetre causes such large alterations in the planet's interactions with all the other planets that variations in its orbit build up and the outcome changes entirely. Since we can never know the positions of all the planets to a centimetre, or, if that accuracy does become possible, to a millimetre, there will always be a limit to the extent of time over which the calculations will work. Since we can never know exactly where to start the calculations, we can never make the calculations realistic.

In modern physics, 'chaos' is the word used to describe behaviour that is predictable in the short term but that in the long term depends so much on the initial state that it cannot be reliably calculated. The weather is an example of chaotic behaviour. It can be predicted, through calculation, more or less accurately, one day or one week ahead. Since no-one could take into account the flapping wings of every butterfly in Brazil at the outset of every calculation, however, meteorologists cannot predict now what the weather will be like in a year's time.

The concept of chaos was discovered as a feature of planetary orbits by the French mathematician Henri Poincaré (1854–1912) in 1887. He was responding to an offer of a prize for the solution to what is known as the 'Three-Body Problem': the exact calculation of the orbits of just three bodies under mutual gravitational attraction. Isaac Newton (1642/3–1727) had given an exact solution to the Two-Body Problem, the orbit of one planet around its sun, for example. The orbit is an ellipse that repeats regularly, and the position of the planet can be calculated to arbitrary accuracy indefinitely far in advance. This is the reverse of chaotic, but it is possible only if there are two bodies (a planet and its sun). By contrast, the solution to the orbits of three bodies proved elusive. Poincaré was able to calculate the orbits numerically – we would nowadays do this by computer, he did it by hand – but he found that the orbits were 'so tangled that I cannot even begin to draw them.' Moreover, Poincaré discovered that when the three bodies began from slightly different initial positions, their orbits could be entirely different. 'It may happen that small differences in the initial positions may lead to enormous differences in the final phenomena. Prediction becomes impossible.' This is chaotic, in the mathematical sense.

As a result of this behaviour, it is impossible to calculate realistically the orbits of a dozen or more planets over a period of time as long as a billion years. What can be done is to make multiple calculations, starting with minor variations in the arrangements of the planets, and see what happens in each case. If you can afford enough calculations and do not get bored by the repetition, the simulations can study a number of theoretically possible planetary systems – more or less like our Solar System – and try to draw out the likely reasons for some general features in the mathematical models that correspond to what we now see of the Solar System in which we actually live.

According to the Nice model of the Solar System, the planetesimals left over from the formation of the planets moved everywhere among the giant planets. Some were ejected from the Solar System as a result of close encounters. As they were ejected, they gave the giant planets a little backward kick, causing these giants to migrate gradually further in towards the Sun. After hundreds of millions of years this brought the two innermost giant planets, Jupiter and Saturn, into resonance, with two of Jupiter's orbits taking exactly the same time as one of Saturn's (see p. 57). This had a profound effect on the other planets. Possibly some of the inner ones were ejected into space or to the Kuiper Belt at this time, leaving behind just the four we know today. Some of the larger Kuiper Belt objects, for example Pluto, originated in this way, having moved from the inner parts of the Solar System to their present icy darkness. There was at that time an alternative future for the Earth, which we can now only imagine, in which it became a distant planet, orbiting in the cold depths of the Solar System beyond Neptune, frozen into a stasis.

Uranus and Neptune were also affected by the repeated, coordinated tug of the two inner giant planets. They moved outwards into more eccentric orbits and ploughed through the vestigial planetesimals left orbiting in the outer Solar System. Most of these planetesimals were expelled from the Solar System. Some did not quite make it into space and were left as members of a slowly moving cloud extending up to a light year from the Sun (the Oort

Cloud) or as an extended disc beyond Neptune (the Kuiper Belt). Others were scattered back down into the inner Solar System. Some of these settled into the region between Jupiter and Mars, and became members of the Main Belt of asteroids. But numerous of the scattered planetesimals collided with each other, the larger planets and their moons.

In particular, the Earth and the Moon were subjected to another bombardment of planetesimals (asteroids and comets as we would now call them), which took place about 4,000 million years ago, after an interval of about 500 million years of relative peace since the first bombardments, which happened during their formation process (including the single impact event of the Big Splash). This second, so-called 'Late Heavy Bombardment' created the majority of the craters still visible on the Moon; also those on the planet Mercury, two of Jupiter's four main satellites and some of the larger satellites of Saturn. It must have created similar numbers of craters on the Earth (and, indeed, Venus and Mars), but they have weathered away, and there remain few or no traces of impacts on these planets from that early time.

The Late Heavy Bombardment heated the Earth's crust to temperatures in excess of 2,000°C. The rocks that had been laid down previously, in the Hadean eon, as the Earth and Moon settled down after the Big Splash, were broken up or pulverized. Both the re-melting and later re-solidification are features of old Moon-rocks 3,800 to 4,100 million years old, found and brought back from the Moon by the Apollo astronauts. The melted surface solidified relatively rapidly as the bombardment died away and the surface cooled.

In the trickle of impacts that took place over the succeeding years, the asteroids and in particular the comets brought water and organic chemicals to the surface of the Earth. The water became the oceans, and the organic chemicals contributed to the soup that was simmered in Darwin's warm little pond (chapter 7). The Late Heavy Bombardment may have set back, or even nullified, any development of life that had previously begun, during the first 500 million years of the Earth's existence, but it brought with it materials that could get the process off to a flying start again. Moreover, it cleared the Solar System of debris so that there was no possibility of a Later Heavy Bombardment (although individual collisions remain a possibility even to this day: see chapter 12).

As a result of the purging of the Solar System of most asteroids by that fortunate resonance between Jupiter and Saturn, further impacts on the Earth are rare enough that they have not caused a devastating extinction of all life; although the residual collisions have certainly had an effect on the development of life, possibly (as we shall see) by drastically changing the global climate and causing past mass extinctions on Earth, even if not life's total demise. It is not known if similar resonances happen routinely in other planetary systems, meaning that they start their development in the same way as our Solar System did, with one major incident in which asteroids and comets were flung everywhere and caused a heavy bombardment of the Earth-like planets at some point but relative peace after that; or whether such resonances are rare, in which case most planetary systems will remain full of orbiting asteroids and their Earth-like planets may be frequently and continuously bombarded. The evolution of life there would thus repeatedly be stopped and restarted. Every time complexity were to build up in the organisms living on such a planet, development would be interrupted and set back near the beginning to start again – a game of Snakes and Ladders with too many snakes. It would be difficult to get from Start to Finish, from archaeon to intelligent life. In the development of life on Earth, this setback happened just once, early on, in the Late Heavy Bombardment, and after this there was time for progressive evolution to reach intelligent life.

After the Solar System was cleared by all these interactions of many of the surplus planets, the remaining planets calmed down to a more orderly existence, by and large settling into orbits that were almost, but not quite, circular. The orbits were slightly elliptical, meaning that at some points in its orbit, a given planet was further from the Sun than at others. The amount by which the Earth's orbit departs from circular is called its 'eccentricity'. The present value of the eccentricity of the Earth's orbit is 0.017, meaning that the longest diameter of its oval-shaped orbit is just 2 per cent longer than its shortest diameter; but this value does change over time and can be much more eccentric, because it still interacts with the orbits of the other remaining planets. Collectively, the planets describe orbits that, in total and over time, intrude into the entirety of the plane of the Solar System, but do not give rise to any risk that they will collide. Thus the Solar System appears to be mostly empty, but in this sense it is exactly 'full'.

* * *

The collisions that cleared out the Solar System until it was 'just full' had a major effect on the evolution of our planet. If the proto-Earth had an atmosphere formed from the solar nebula, this was probably mostly made of the lighter gases, such as hydrogen and helium, and the atmosphere was ejected into space in the giant impact with Theia. The heat of the collision – heat created from the accretion of smaller planetesimals both at the formation of the proto-Earth and at the Late Heavy Bombardment – and radioactivity all resulted in much of the Earth's surface becoming molten. Steam was produced from the minerals in the crust of the Earth, and steam and yet other gases were released by volcanoes. Asteroids and comets brought further amounts of water and other gases to the Earth. These gases formed the first oceans and a second atmosphere, which evolved into the one that remains with us today. The first oceans existed by the end of the Hadean eon, as evidenced by the discovery of old 'pillow basalts'; when volcanic rocks solidify in lakes or seas, they form in blobs that look like pillows, the shape of which gives away the fact that they formed under water. Similarly, lakes and oceans existed on Mars at this time (see chapter 13).

The second atmosphere was probably made of gases similar to those created by modern volcanoes: water (H_2O), carbon dioxide (CO_2), sulphur dioxide (SO_2), carbon monoxide (CO), sulphur (S_2), chlorine (Cl_2), nitrogen (N_2), hydrogen (H_2), ammonia (NH_3) and methane (CH_4). This was the atmosphere that, in Stanley Miller's experiments (see chapter 7), produced amino acids and apparently seeded life, perhaps deep in the oceans. There was, however, a factor that inhibited the development of life, except in the very deep parts of the sea: the deleterious effect of ultraviolet light on living organisms. Until life itself reduced the amount of ultraviolet light that reached the Earth's surface, the copious amounts emitted by the Sun restricted the development of life in the ocean shallows, in lakes and on land.

The term 'ultraviolet' means simply 'beyond the violet' in the spectrum of sunlight, and the effect of ultraviolet radiation varies with how far 'beyond the violet' this light is. Ultraviolet radiation at longer wavelengths, those just beyond the violet, with wavelengths of 320–400 nm,[4] is like a colour, but has only a scientific name, UV-A. From an anthropocentric point of view, UV-A is both helpful and harmful. We use it to form vitamin D in our skin. Too much, however, and UV-A causes sunburn on human skin and cataracts in eyes. Further away from the violet, at even shorter wavelengths, from say 280 to 320 nm, lies UV-B. UV-B causes damage to living organisms at the molecular level. Its effect is complex, but in humans UV-B induces skin cancer and suppresses the body's immune response to various viruses. In many land plants and in phytoplankton in water, UV-B reduces the effectiveness of photosynthesis, as well as increasing susceptibility to disease. In the Antarctic Ocean, increased exposure to UV-B radiation reduces the effectiveness of photosynthesis by phytoplankton in surface waters by 10 per cent. UV-C is the most extreme form of ultraviolet light, covering the wavelength range 100–280 nm. It is the most dangerous type of UV radiation, directly breaking apart organic molecules.

What prevented large amounts of ultraviolet radiation reaching the surface of the Earth? Indeed, what changed the make-up of the atmosphere of the Earth from its Hadean composition to that of the present day? Some of those gases of the second atmosphere are highly reactive chemicals, and they combined with other chemicals in rocks. The crucial change was that oxygen was released into the atmosphere and combined with other atmospheric gases. Today the atmosphere of the Earth is 78 per cent nitrogen (N_2), 21 per cent oxygen (O_2), 0.03 per cent carbon dioxide (CO_2), plus traces of other gases (such as water, H_2O). But where did the oxygen come from?

Some of the oxygen came from inorganic processes: the break-up of water and carbon dioxide molecules by ultraviolet light from the Sun. The water molecules broke up as follows: $H_2O \rightarrow H + OH$. The two components that were produced this way are called 'radicals' – fragments of a molecule, floating about with a great desire to recombine chemically with something. The two radicals produced were a hydrogen atom (H) and a hydroxyl radical, made of a hydrogen atom and and oxygen atom (OH). Likewise, the break-up of carbon dioxide molecules, $CO_2 \rightarrow CO + O$, produced carbon monoxide (CO) and an atom of oxygen (O). The OH is very reactive and combined with the O to make molecular oxygen and hydrogen atoms: $O + OH \rightarrow O_2 + H$. The hydrogen atoms are light and readily escaped to space, allowing the oxygen to build up in the air. This started to happen when the Earth was about 2,000 million years old, and it is the start of what is known as the Great Oxygenation Event, one of the major episodes that enabled life to flourish on Earth. This Event was the birth of Gaia.

The build-up in concentration of oxygen started slowly because, at the same time that it was made, large quantities of the highly reactive gas were absorbed in chemical reactions with minerals dissolved in seawater. At first, only a small amount of oxygen remained in the atmosphere, at a concentration that was less than 1 per cent. The crucial property, even of this small concentration of oxygen, was that it formed some ozone. Ultraviolet light can split an oxygen molecule (O_2) into two oxygen atoms ($O + O$). An individual oxygen atom (O) may combine with an oxygen molecule (O_2) to form a molecule with three oxygen atoms, ozone (O_3). This gas is present in the atmosphere now in a layer at an altitude of 10 to 50 km. Ozone serves now, as it began to do then, to shield the Earth's surface from ultraviolet radiated from the Sun into the Earth's atmosphere; the Sun's ultraviolet light created its own sunblock as ozone in the Earth's atmosphere.

*   *   *

With the damaging UV sunlight much reduced in intensity, cyanobacteria and more complex plants developed in shallow water and, later, on solid land. They benefited from the brightness of the sunlight in shallows near the shores of the oceans. In the chemical reaction of photosynthesis, cyanobacteria and plants take in carbon dioxide and, using sunlight, combine it with water to make oxygen. This accelerated oxygen production, but for a long time it remained at a low concentration in the air, since it continued to be taken up by seawater and then by solid rocks on dry land. This is recorded in rocks known as Banded Iron Formations (or BIFs). These are ancient sedimentary rocks made up of repeated thin layers of iron oxides, separated by iron-poor deposits of such other sedimentary rocks as shale and chert.[5] It is not known why the iron oxides are banded – perhaps the deposits were seasonal, or perhaps the amount of oxygen that was produced by the atmosphere oscillated for some unknown reason. Although the take-up of oxygen by rocks still happens, oxygen began to be produced by plants at a rate faster than rocks can absorb it, and eventually, only about 500 million years ago, the oxygen level in the atmosphere reached about 20 per cent, perhaps even 30 per cent at some stages. It is no coincidence that this period saw the beginning of the proliferation of life on Earth in the Cambrian period.

Once life has taken hold on a planet with an oxygen atmosphere, protected from damaging ultraviolet radiation from the Sun, it might be able to adapt to changes that decrease the sunblocking effect – loss of atmosphere by leakage into space, for example. This scenario is relevant to the planet Mars, which once had a denser atmosphere but has since lost it (chapter 13).

One example of this happening to life-forms from our own planet was investigated in 2008–10 by members of a Spanish team from the Instituto Nacional de Técnica Aeroespacial in Torrejón, Madrid, working in collaboration with scientists from the Open University in the UK and from the Institute of Aerospace Medicine, Cologne. Lichens are plants formed from the symbiotic association of certain fungi and algae or cyanobacteria. Samples of rocks with lichens were collected from coastal limestone cliffs in Devon, in situations where they were exposed to periodic desiccation, saltwater (during high tide), freshwater (during rain at low tide), temperature fluctuations, solar radiation and the generally nutrient-poor environment of the limestone surface. They were thus a community of extremophiles, some of which might be capable of tolerating the conditions associated with outer space, including complete lack of shielding from solar ultraviolet radiation. In two experiments, the rocks were launched into orbit as part of the European Space Agency's Biopan VI mission on the Russian Space Agency's Foton capsule, and in STS-122, a NASA

Space Shuttle mission to the exterior of ESA's Columbus module on the International Space Station. The lichens stayed in space for 10 and 548 days respectively. Numbers of the micro-organisms survived the cosmic conditions, which acted as a selective pressure on the communities, favouring some colonies of micro-organisms and killing others.

Lichens in the Antarctic are also subject to environmental extremes, including not only low temperatures, but also desiccation and ultraviolet radiation at higher levels than elsewhere on Earth. Air at the poles is very dry, because the global circulation of air is driven by the overhead heat of the Sun at the equator causing air there to rise, an effect that then draws air from the poles, which is replaced by cold dry air that falls from the stratosphere. In Antarctica the air is especially dry because the Antarctic landmass forms a plateau at a high altitude, so the air is characteristically that of a high mountain. UV radiation over Antarctica is also a strong stress; UV is intense because the concentration of ozone in the atmosphere above Antarctica has been reduced – the so-called 'hole in the ozone layer'.

In response to their environmental conditions, lichens make chemical compounds to protect against stress. The compounds contribute to the colour of the organisms. Some lichens in the Antarctic colonize rocks. One type, collected by the British Antarctic Survey, lives just under the surface of rocks and produces an orange pigment, beta-carotene (the pigment that makes carrots orange), which is a UV-radiation protectant, at the exposed upper surface of the lichen. This interface also contains whewellite, a mineral produced by the lichen that seals it against desiccation. It is possible that similar defensive mechanisms might be produced by life on Mars – if there is any – in response to changes in climate.

The availability of free oxygen in the atmosphere enabled the evolution of oxygen-breathing life. The energy that creatures can produce in this way enables them to grow bigger, react promptly, mobilize quickly: in general, to act in complex and interesting ways. The Great Oxygenation Event, the change of the Earth's atmosphere to one that is oxygen-rich, was a key factor in the successful development of intelligent life and, for better or worse, its capability to exercise control of our planet.

* * *

It seems as if, given a commonly available early atmosphere, a rocky planet in the habitable zone will make life that, if Earth-like, will in turn cause the atmosphere to evolve to contain oxygen, which will amplify at least temporarily the life-bearing possibilities of the planet. If we find a planet with an oxygen atmosphere, this will be a good indication that it is likely also to be the habitat of plant life that converts carbon dioxide to oxygen. As we have seen, the availability of oxygen additionally favours the development of life with a more effective metabolism and therefore greater capability, including possibly greater intelligence.

This syllogism has been a principle on which space missions have been designed to search for life-bearing planets. The starlight reflected from the surface of such a planet, or the infrared radiation that emanates from its warm rocky surface, will pass through the planet's atmosphere. Oxygen, in its most common form as a molecule of two oxygen atoms (O_2), or as a molecule with three, ozone (O_3), will leave its traces on the spectrum of that light, absorbing key wavelengths that are distinct proof of that gas and no other. Both NASA's Terrestrial Planet Finder and ESA's Darwin are space missions that include this as a feature of their design. They are predicated on the use of advanced optical techniques to separate the radiation that comes from the planet from that from its parent star, to make an image or an equivalent of the planet itself. Depending on the success of the technique and the brightness of the planet, the mission could go on to study the characteristics of the radiation and detect oxygen. Alas, at present neither project has got beyond the stage of design studies, because the technology is considered too challenging and therefore, in practice, likely to be too expensive for space-science budgets to bear.

At the moment, therefore, we are not in a position to find a planet with an oxygen atmosphere on which we can focus our efforts to find life in the Cosmos. Given some technological breakthroughs and more affluent times, we will be able to set out on that search, hopefully in the not-too-distant future.

9

The Evolution of Life

It takes a long time for life to emerge, and even longer for it to evolve to produce complex organisms. Life, therefore, is not likely to be found on young planets, and on any older planet on which favourable conditions began only recently it will not yet have evolved to become complex. What does this tell us about the likelihood that we are not alone? How will we recognize extraterrestrial life when we see it? It is difficult even to compile a definition of life here on Earth that lists its distinctive characteristics, while excluding what we would term inanimate. But advanced life has, in general, the following properties.

Living organisms sense and respond to external stimuli. This, as far as the Greek philosopher Aristotle (fourth century BC) was concerned, was the defining characteristic of life. Plants respond to external cycles of day and night, summer and winter. Animals sense the external environment and alter their behaviour, whether that is regulating body temperature, moving, seeking food or kicking a football.

Living organisms also respond to internal stimuli. These arise inside the body and the body reacts to them; for example, in a human being, if blood-sugar levels rise, insulin is released, which then causes the concentration to decrease. Through such changes, which are often brought on by regulating and diversifying metabolic processes, living organisms stabilize their internal environment: a condition called homeostasis.

Sensing the environment and controlling movement in response to it is a complicated process, and it is the main reason that some living organisms have brains. Possessing a brain is a rare attribute; 80 per cent of living organisms do not have one. The brain is a central processor that makes sense of input information, especially sight, which puts a great demand on processing power to infer shape, texture, colour, three-dimensionality and movement, and to combine them into recognition and reaction, often very quickly – within milliseconds. Our ancestors developed the ability to see tigers in the grass and run away. Even relatively small creatures with small brains have this ability; it is hard to catch a fly because it sees your hands approaching and rapidly veers away. The brain is also a controller, moving such simple parts as levers (bones) and strings (tendons) that are connected in complex structures, causing them to move in timed sequence with exquisite precision, in close relationship to the brain's intent.

Is it possible for an extraterrestrial creature to have no brain as such and to rely on distributed intelligence, thinking, not with one organ, but with the whole body? Our brains are specialized organs, and they make considerable demands on our bodily function; for example, the brain must be kept at a much more constant temperature than the rest of us. In extreme cold, the temperature controllers within our bodies will withdraw heat from less important, less heat-efficient parts in order to maintain brain temperature. Mountaineers get frostbite in toes, fingers, legs and arms: parts that are sacrificed to keep their brains alive for longer. It seems likely that extraterrestrial intelligence will reside in a similar, specialist, isolated organ.

Extraterrestrials will have evolved sensory organs in order to be able to respond to and manipulate their environment. A limited range of senses is likely, because of the limited range of possible input stimuli. The sense of touch is probably universal, since every organism will be in contact with solid surroundings and touch will enable the organism to optimize its position. Smell and taste are senses that depend on molecules being brought to an organism's surface, by the wind or by physical transference, and enable the organism to optimize its position over a wider area, to gather food, for example.

Electromagnetic radiation is common on a planet near a sun. Our Sun produces its main radiation output in the visible part of the spectrum, and the common atmospheric gases, such as nitrogen and carbon dioxide, are transparent to light. Sight is such a useful sense for a complex organism that on Earth eyes have developed independently several times, and descended in distinct lines of species. Other planets will have similar properties, and their extraterrestrials might well have eyes. If so they will be placed in an advantageous position to enable the life-form to see – for example, up high, and they will be very near to image-processing equipment, i.e. a brain. Indeed, the demands of vision on the capacity of the brain have been a major reason for its evolutionary development. Such extraterrestrials may well have two eyes, since there is an advantage in stereoscopic vision, but may well have more – enabling them to see both in front and behind, as well as above. The temperatures on a habitable planet will be such that there will be abundant and useful infrared radiation coming from important things in the environment.

All warm things radiate infrared radiation: it is the radiation that transmits heat. Hotter things radiate more infrared than colder things, so infrared vision could be useful to identify features of the environment, like food or predators, as some reptiles, such as snakes, are able to do. On the other hand, little or no X-radiation is produced on the surface of a planet, and X-rays are absorbed by air, so there would be no evolutionary push for extraterrestrial life to have X-ray eyes on an Earth-like planet.

Sound is the result of motion and vibration, which occur everywhere on the surface of a planet. Ears would be useful to an extraterrestrial, and it would be an advantage to have a wider frequency range than we do. Humans hear frequencies between 20 Hz (cycles per second) and 20 kHz (20,000 Hz). Other mammals that live on dry land may hear much lower frequencies (down to 10 Hz) or higher, up towards 100 kHz. Mammals that live in water, for example dolphins, hear frequencies up to 150 kHz. The ability to hear an extended frequency range enables faculties that have not been developed in humans, such as echo-location in bats and dolphins, or the ability to communicate over large distances, as elephants and whales do.

*   *   *

Living organisms grow. Living organisms use energy: taking it in, modifying it for use, using it to regulate their internal environment and releasing the by-products. A living organism's parts all increase in size over its lifetime, usually more quickly in youth than in the adult stages. A living organism often develops in appearance and function during its life: frogs grow from small eggs to larger tadpoles to adults that are larger still; ducklings are small, fluffy and flightless, and grow into larger adults with feathers that are streamlined and enable the duck to fly. This is called ontogeny. But organisms grow only for a finite amount of time. Eventually an organism dies and its life processes cease.

Plants take in sunlight and use its energy to synthesize sugars from the carbon dioxide in the atmosphere, excreting oxygen. Archaea, on the other hand, produce methane, by breaking down organic material in the absence of oxygen. The presence of both gases could be highly significant for the search for extraterrestrial life. If they exist in the atmosphere of a planet, the atmosphere must be out of equilibrium – something must have just produced the gases, because as soon as they are made they begin to combine with other chemicals and will not exist in their separate forms for much longer. Terrestrial Planet Finder and Darwin, space projects proposed by NASA and ESA, target the detection of oxygen in an Earth-like planet as a sign of life, while the presence of methane in Mars's atmosphere has been acclaimed as a possible sign that there is residual life on the Red Planet. Methane-producing archaea on Earth are both free-living organisms and organisms that live in association with other creatures, for example in the guts of ruminant animals, such as cows. These animals release methane by passing wind. Methane is thus a signal that life exists on a planet.

* * *

Living organisms are complex. Life is organized into groupings of greater complexity; for the eighteenth-century German philosopher Immanuel Kant, this was the most important property of life. Starting with atoms and molecules, life is assembled into cells, tissues and organs, a working collection of which constitutes an individual in a population of a species. Species interact and assemble into an ecosystem.

atoms → molecules → cells → tissues → organs → individuals → populations → species → communities → ecosystems

In the ordinary course of events, we readily observe ecosystems; communities; the various species, populations and individuals of which they are composed; and some of the external organs of which they are made. We would need a laboratory to penetrate below this level and perceive the biochemical and biophysical groupings of life beneath the surface of the life-forms that we see.

At each level up in the hierarchy of organization, new properties emerge from the assemblage of the smaller parts, bringing new capabilities: at each stage, the whole is greater than the sum of the parts. Because there are a large number of components that assemble into each stage, there are very many possible arrangements or groupings. A biochemical molecule may be as few as half a dozen but as many as thousands of atoms. Atoms are 1/10,000th of a micron in size, cells are typically 10 microns in size, and an animal might be typically 10 cm to 1 m in size. There are 1,000 million million atoms in a cell, and a similar number of cells in a living organism. Bacteria are very simple single-celled organisms, but each cell contains up to 1,000 kinds of molecules. In a word, this complexity is the reason for biodiversity: the extraordinary range of living organisms that exist in our world, and, possibly, in others.

* * *

Living organisms reproduce themselves. Plants and trees create seeds from which grow replicas of their parents – not exact replicas, because of the mixing of genetic material from males and females, and because of accidental alterations, but recognizably similar creatures. They do this by creating flowers, fertilized by pollen from other members of the same species brought there by insects, by the wind or otherwise. They also reproduce themselves by propagating from shoots or bulbs underground, which grow into more precise replicas of the parent plant. Insects, such as dragonflies, lay eggs from which spring fully formed adults. Frogs create eggs that develop into tadpoles and then into adult frogs. Birds, such as doves or ducks, lay eggs with hard shells from which emerge small versions of birds, which develop over a few weeks into creatures like their parents. Animals – dogs and donkeys, for example – reproduce by giving birth directly to small immature versions of themselves, which over a longer time, months or years, develop into adults. The reproduction process of a species of living organism is the means by which the species survives and develops, even though each given individual dies. If all the individuals in a species die, the species itself ceases to exist. We say then that the species becomes extinct.

What this means in the context of the development of life on other planets is that once life starts in such a situation, it is likely to persist until conditions there change drastically.

* * *

Living organisms evolve. When living organisms reproduce, they mutate and evolve through natural selection. This is such a distinctive property that many biologists consider it the single most important thing that distinguishes life from the inanimate. Evolution occurs because the reproduction process has built-in mechanisms to select characteristics that come from the parents and mix them up in different combinations. The process is also error-prone to some degree, and can result in a living organism with slightly different properties and capabilities than its ancestors: a new combination of genes or a mutation. The reproduction process is mostly stable, so once it has been created, the new combination can persist and may become more frequent if it has some advantage that results in increased reproductive success, if for example it makes the creature better than its fellows at feeding, hiding from predators, resisting illness or mating. This will increase the proportion of the species with the favourable variation and intensify it; eventually the new creatures will become so distinct from their predecessors that they become a new species. This is the process of 'natural selection', identified by Charles Darwin in 1859 in his book *On the Origin of Species*.

Evolution and natural selection are, in part, accidental, because new combinations occur at random and because the environment of any given species can change unexpectedly. On the other hand they are somewhat predictable, taking place within some prior constraints. Sight is of such clear advantage that eyes have evolved independently, and repeatedly, in the origin of new species. Mobility is another obvious advantage, and creatures have evolved many different ways of getting around: walking, slithering, flying, propelling themselves in water, and so on.

In order to manipulate its environment, to farm food, make shelter and extend capability by the production of tools, a life-form needs an efficient metabolism so that it does not have to stop constantly to eat, and efficient manipulative appendages – arms and hands – perhaps evolving from the limbs with which the organism moves.

* * *

In terms of what we can expect, if and when, in our future interstellar explorations, we encounter advanced extraterrestrial aliens, it could be one of two things. On the one hand the aliens could be just that – alien. In the words of the palaeontologist Simon Conway-Morris (b. 1951), they could be 'constructions so unfamiliar that they are only brought home by accident and then inadvertently handed over for curation in a department of mineralogy.' On the other hand, he says:

> Far from blinkering our outlook, our local biology tells us all we need to know.... No sentient forms weaving their existence in vast interstellar dust clouds, farewell to bizarre filamentous species greedily soaking up the intense magnetic fields of a crushingly oppressive neutron star and, even on Earth-like planets, no forms that we might as well call conceptualised pancakes.

We have an image in our minds of extraterrestrial aliens, which is dominated by what we see in popular culture, in magazines and films. For a clear reason, the aliens in such films are most often humanoid in form: bipedal, with arms and hands, a head (or two), eyes, a mouth with which to eat, and usually comprising two genders. The obvious reason is that film directors dress up human actors to play the part of extraterrestrials, and there is only so far that they can economically go to disguise the basic structure of a human being in a costume. Because his budget was running out, George Lucas had to use a relatively limited range of masks and costumes to depict the assorted extraterrestrials in the Wookiee Chalmun's canteen on the planet Tatooine, in the film Star Wars; this is the reason why the bar's customers are nearly all humanoid.

This fiction has been coloured by reports from people who claim to have encountered extraterrestrial aliens. Perhaps the most notorious set of reports comes from the Roswell incident, which occurred near Roswell, New Mexico. Roswell is a ranching town, the place near which, in the 1930s, the space pioneer Robert H. Goddard (1882–1945) carried out his experiments, pioneering the design of liquid-propelled space rockets that laid the foundation of present-day space programmes. Goddard's work is featured in the municipal Roswell Museum and Art Center, where his early workshop is reconstructed. But Roswell is more widely known, and is an attraction for tourists, through its International UFO Museum, a converted cinema on Main Street. Here the Roswell incident is described through newspaper reports, affidavits, pictures, models and maps.

According to the sometimes contradictory accounts, on 4 July 1947, witnesses near Roswell saw and heard what they assumed at the time was an aeroplane crash. The next day, one witness, a rancher, came across a debris-field of 'stuff', such as aluminium struts (with marks on that were said to look like writing in an unknown alphabet) and metal foil. He took a sackful to Roswell to show to the sheriff. Following a spate earlier in the year of reports of sightings of flying saucers, the local newspaper described the crash as of a flying saucer or disc. There was an airforce base nearby. Military personnel from the base became involved, and eventually described the crash as that of an experimental high-level weather balloon, but the talk in Roswell at the time and since is that they swapped the real debris of an interstellar spacecraft for innocuous material and attempted a cover-up. All this has become the subject of a long-running conspiracy theory, human interest in which has not diminished as time passes.

Various people have reported that a number of small humanoid bodies were taken from the site and subjected to autopsy. There is a report that one unnamed witness, a nurse, participated in the examinations; she is said to have drawn what is presumed to have been the extraterrestrial crew of the spacecraft. They were described as small, childlike in size, with big eyes. In the International UFO Museum, they are variously shown in pictures and as models, apparently naked, standing erect. One representation of them is situated in front of a flying saucer that from time to time emits dry ice fumes and plays ethereal 'Ooo-ooo-ooo' music. The aliens are four-limbed (two arms, two legs), with a head and face with two eyes, a nose and a mouth, all supported by a neck, a backbone and a ribcage. Apart from having four digits per limb rather than five, and big eyes and no genitals, they appear very like human beings. They are unthreatening, indeed rather appealing, and you can scarcely buy anything in Roswell that does not bear a picture of one of them. You can also eat in a flying-saucer-shaped McDonald's restaurant, perhaps to celebrate Roswell's claim to interstellar fame on its Extraterrestrial Culture Day every year on the second Tuesday of February.

The various versions of the story are riddled with inconsistencies, and some of the accounts have been clearly exposed as hoaxes. The fiction of films and of the Roswell crash might, however, be close to reality; evolution may push the development of advanced life towards familiar architectures, and, if and when we encounter aliens with whom we can communicate, they may indeed be somewhat like us.

Of course evolution did not end when it created humans. We can see signs of continuing development in the withering away of such human organs as the appendix, which now has only a residual use in the immune system, and which we can do without. What lies on the evolutionary path beyond human-oid intelligent life like us? Our first indications may come from encounters with an extraterrestrial alien with a longer evolutionary history than ours.

* * *

The reproductive and evolutionary process is carried out in the vast majority of species through the famous DNA (deoxyribonucleic acid) molecule, the struc-ture of which was identified, in a Nobel-prize-winning discovery in 1953 by Francis Crick (1916–2004) and James Watson (b. 1928), to be a 'double helix'. The molecule is made up of two long strands that are linked together in a helix shape, side by side, by bonds that run from one strand to the other like the rungs of a ladder. Each strand consists of a sugar-phosphate backbone, with, mounted inside and facing the other strand, a sequence of only four kinds of chemical compounds called bases: adenine, thymine, guanine and cytosine (usually abbreviated as A, T, G and C). The bases are linked across from one another in pairs, because the adenine–thymine bond is exactly as long as the cytosine–guanine bond. So, in a long ladder-like structure, adenine can join only with thymine, and cytosine can join only with guanine. Each rung of the twisted ladder in the helix is thus of equal length, and the sugar-phosphate backbone is smooth, neither under strain from internal stresses nor kinked. The DNA molecule is large, complex and stable.

The helix can unzip down the middle, and the freed links can attract more bases, with each half-ladder acting as the foundation on one side for a com-plete double helix. Since only one base fits each link, the DNA molecule reproduces itself exactly – unless it has been altered by any of a number of causes. In a famous understatement, Watson and Crick wrote in the scientific paper published in *Nature* on 25 April 1953, in which they announced the structure of DNA, that 'It has not escaped our notice that the specific pairing we have postulated immediately suggests a possible copying mechanism for the genetic material.'

The rate of evolution of an organism depends on the rate of reproduction of the DNA molecule that it hosts, and the rate at which recombination of genes or mistakes occurs. The DNA molecule can spontaneously repair some mistakes, if there are few of them. Most changes are presumably dead ends – perhaps the organism does not survive at all, perhaps the consequences of the mistake are immaterial to its survival, or perhaps they are deleterious. In those rare cases where a favourable mistake is made, and chance delivers the organism the opportunity to reproduce in a long-enough lineage, it can progress from one species into another. The essential ingredients of evolution are thus not only a method of reproduction that allows for recombinations in a way that is not entirely perfect, but also time. We are more likely to find complex extraterrestrial life in an older world than in a new one.

* * *

The first living organisms were simply cells within a wall. Everything the wall contained was mixed up in a structureless way, inside, including their DNA. The biological name for such organisms, which usually consist of a single cell, is prokaryotes. Bacteria and archaea are prokaryotes. Unlike the cells of eukaryotes, the cells of prokaryotes have no nucleus. Eukaryotes do have a nucleus, and their DNA is inside this central structure, enclosed by an internal membrane. As outlined above, the first prokaryotes emerged on Earth at the end of the Hadean eon, about 1,000 million years after our planet was formed. They evolved into various types of archaea and bacteria, such as the cyanobacteria that formed stromatolites. Eukaryotes probably emerged when the Earth was aged between 1,800 and 2,900 million years. By the end of that time, eukaryotes consisted of approximately ten cell types. Prokaryotes, like bacteria, associated with one another and evolved together into larger single-celled eukaryotes, with more complex structures and internal divisions, such as mitochondria, hydrogenosomes and chloroplasts.[1] This produced a much greater variety of complex multicellular eukaryotes, including green and red algae, and amoebae.

* * *

The history of the evolution of life on Earth was, for the first 3,900 million years of the Earth's existence, purely a history of single-celled organisms, with the most complex structures being associations of colonies of such organisms, such as the bacterial mats that form stromatolites. Such associations did not reproduce en bloc or remain in fixed relationships – one colony with another – as a biological species does. When this property did emerge, multicellular organisms appeared. These are characterized by having different cells in fixed relationships that carry out various functions, all of them necessary for the continued proper survival of the species. This seems to be a natural thing to happen, and one that confers a significant reproductive advantage, because it was so successful. It may even be that the emergence of multicellular organisms occurred independently on more than one occasion. Such organisms first appeared in the sea about 600 million years ago, during the Ediacaran period, at the end of the Neoproterozoic era, just before the Cambrian.

The first multicellular species were relatively simple, and sometimes unrecognizable: such organisms as sponges, algae and slime moulds. These were, however, the precursors to the evolution, in the Cambrian explosion, of early animals and large plants (500 million years ago). Other animals quickly followed, first reptiles and amphibians (300 million years ago), and then, branching from reptiles, mammals (200 million years ago) and birds (100 million years ago).

It was primarily after the Cambrian explosion that life existed on Earth in such quantities that rocks were created from biological material. Coal is rock that is mainly carbon produced from plant material that accumulated under water, with air excluded so that the carbon did not oxidize to carbon dioxide. Overlaid by further strata of limestone, shale and sandstone, the plant material was compressed and fused into solid rock, in layers called 'coal seams'. The main time at which this happened is named after coal itself: the Carboniferous period, lasting from 359 to 299 million years ago. In other circumstances, tiny plankton accumulated in large quantities at the bottom of the ocean, and the pressure of more strata laid on top caused oil and natural gas (such as ethane and methane) to be squeezed from carbon of biological origin in the shale, and perhaps trapped in reservoirs in impervious layers. Chalk is a white sedimentary rock, formed under the sea by accumulations of calcite from marine micro-organisms. It is often found in association with flint, the nodules of which formed in holes in the chalk from the accumulation of a viscous substance made from silicaceous sponges and micro-organisms. Chalk deposits derive mainly from the Cretaceous period, 145–66 million years ago. Coral is even more obviously biological: deposits were created directly, as reefs, from the bony structures of living polyps.

Although such massive recent strata as coal or chalk are obvious biological rocks, traces of biologically produced minerals are found in some pre-
Cambrian rocks – the residue, presumably, from bacterial colonies.

* * *

Apart from finding fossils of once-living organisms in a dead rock, there is an
interesting way to determine whether or not it is made from material that has
been processed biologically. Chemical elements are made of atoms that can
exist in slightly different forms. As already explained (on pp. 51–52), atoms are
made of a cloud of electrons (which are very light) that orbit a heavy nucleus.
As well as protons, nuclei contain neutrons. All atoms of a given element have
the same number of electrons, and an exactly equal number of protons. Thus
all oxygen atoms have eight electrons in orbit around eight protons.

The nuclei of oxygen atoms, however, can possess varying numbers of
neutrons. Eight, nine or ten neutrons are typical; sometimes oxygen nuclei
exist with more or less, but they are unstable and decay by radioactivity into
something else. Atoms with varying numbers of protons and neutrons are
called isotopes, symbolized as the usual letter(s) that abbreviate the chemical
name, labelled beforehand with a superscript representing the number of
protons plus neutrons in the nucleus of the isotope. The isotopes of oxygen
are thus symbolized as ^{16}O, ^{17}O and ^{18}O. Most elements of biological interest
(including carbon, C, hydrogen, H, oxygen, O, nitrogen, N, and sulphur, S) have
more than one stable isotope. Isotopes of the same element take part in the
same chemical reactions, so biological chemical processes affect all the atoms
of a given element equally, although the processes take place in different
physical situations. For example, a chemical compound might be made in a
process that involves the evaporation of some gas, or it may require a chemical
compound to be pushed in a solution through a membrane, for example a cell
wall. But because the atoms of different isotopes are of varying weights, they
react at different rates: the physical processes involved in chemistry discrimi-
nate between heavy isotopes and light ones, even if the chemical processes
themselves do not. As a result, chemical processes, whether taking place in
inorganic circumstances or in biological material, can result in reaction
products that are a little bit heavier or lighter than usual.

Measurements of the proportion of some isotopes in material can therefore be a clue as to how it has been physically, chemically or biochemically
processed. One important application of this technique is to determine from
where in the Solar System material originated. Because the heat of the Sun
preferentially evaporated lighter ices from the innermost parts of the Solar

System, isotopic ratios vary from planet to planet. This is one way that the origin of meteorites is determined. On the ground, a meteorite is a rock that looks like many others; in the right laboratory the isotopic ratios can reveal that the rock is from Mars.

Likewise, life can leave traces in the isotopic ratios of its involvement in the history of inert material – the carbon isotope composition in carbonate deposited by bacteria is different from carbonate material made by an inorganic process. The technique of isotope analysis can identify whether life once existed in a now-dead world. This is likely to be the way in which methane found in Mars's atmosphere will be determined to be from living creatures or from volcanoes. It could in general be the way in which the existence of life is first proved by robotic spacecraft, sent to explore possible places in the Solar System where life might be or have been (chapters 13–15).

* * *

If the history of the evolution of life on Earth is typical of that on any similar planet, in a similar planetary system, the simplest conclusion is that the chemical ingredients for life are available from the birth of the planet. The planet takes some time to become stable enough for life to take hold, but simple, unicellular forms appear fairly quickly (on a geological timescale) once the physical conditions are right, within, say, 1 billion years. Unicellular life develops for a further 3 billion years into increasingly complicated forms, with even more complex, multicellular life appearing 4 billion years after the planet's birth.

It seems that it is almost routine for a planet to accumulate and make organic chemicals, the ingredients for life. Given some time, the planet could well settle down to a quiescent-enough state to provide the right environment in which, if life is created, it survives. It appears (from Stanley Miller's experiment, see chapter 7) that it is relatively easy to make simple life-forms. We have a good chance of finding archaea and bacteria elsewhere on other planets. But it is much harder for complicated life-forms to be generated. The most important condition is that the right environment must remain stable for a sufficient length of time for evolution to progress.

The key ingredient to enable advanced life to develop is therefore not the chemical mixture, the planet, nor its place in the planetary system: but simply enough time.

10

The Sun: Source of Vital –
or Death-dealing – Energy

The Sun is a star. Like all planetary systems, our Solar System depends on its star for its existence, stability and, at least in the inner zones, for warmth. Free-flying planets are feasible, and may indeed be very numerous, but, far from their parent star, they would be extremely cold places where life would struggle to survive, or even to begin. Stars support the main ecosystems in the Cosmos in which life is most readily possible.

The Sun exists in space as a mass of gas, principally hydrogen, which has contracted from a vast interstellar cloud of hydrogen atoms and arrived at a state in which it supports itself against its own force of gravity. The downward thrust has compressed the internal gases of the Sun and increased the density and pressure within. This compression has also heated the interior. The combination of density and heat means that the hydrogen atoms now collide with such force that they split to become electrons and protons. The protons crash together and, through nuclear reactions, fuse into helium nuclei. This releases nuclear energy, which heats the Sun's material even more. In this way, the Sun creates enough internal, outwards-thrusting pressure to support its own enormous weight, and at the same time provides the radiation that warms our planet.

The Sun is highly stable because it is self-regulating. If, for some reason, it gets bigger, the pressure within reduces and the Sun settles back to its former size. On the other hand, if it gets smaller, the pressure and the density within increase, more energy is generated, and the Sun puffs up again. It maintains its broad characteristics over long periods. Not all stars are like this: some are variable, and change size sporadically or regularly, in hours, days or years. When this happens, they also change in temperature. The amount of energy a sun radiates depends on its surface area and temperature, so if it changes in size, any planets in proximity will become warmer or colder, accordingly. Life would likely be impossible on a planet with such a variable cosmic environment.

The Sun radiates a very broad spectrum of energy, primarily in the infrared, visible and ultraviolet wavelengths, but extending from radio wavelengths to energetic X- and gamma-rays. This energy warms the Earth to a temperature that is hospitable to life.

About 34 per cent of the Sun's energy that reaches Earth is reflected back into space from the tops of clouds, ice and snow. Of the radiation intercepted by the Earth, 66 per cent is absorbed in one way or another, heats our planet, and is re-radiated back into space. The balance between the radiation that falls on Earth and that which is reflected back into space is what maintains the temperature of our planet. Changes in cloud or ice cover, and even, in recent times, aircraft condensation trails, affect this delicate balance.

* * *

As well as maintaining Earth at a habitable temperature and keeping most of the water here in liquid form, the Sun directly powers life-giving natural processes. Earth's energy sources are, ultimately, all astronomical (**Table 5**), and

Table 5. The Astronomical Sources of Energy on Earth

Energy			Effect and location
Energy sources from the Big Bang			
Hydrogen fusion energy			Energy of the Sun
Hydrogen fusion energy			Fusion reactors
Accretion energy			Geothermal energy from residual warmth in the Earth
Energy sources from ancient stars			
Atomic nuclei →	Nuclear energy		Nuclear reactors
Atomic nuclei →	Radioactive heating		Geothermal energy from the Earth's core
Solar radiation			
Light and infrared			Climate of the Earth
Light and infrared			Solar power at Earth's surface
Evaporation and rain			Hydropower
Wind			Wind power
Wind →	Waves		Wave power
Photosynthesis →	Plants		Vegetable biomass for food, to burn, etc.
Photosynthesis →	Plants →	Animals	Animal biomass for food
Photosynthesis →	Plants →	Vegetable biomass	Wood to burn
Photosynthesis →	Plants →	Coal, gas and oil	Fossil fuels to burn
Energy sources from the Moon			
Orbital energy →	Oceanic tides		Tidal power

the Sun is responsible for more of them and much, if not most, of the energy generated by them than any other astronomical phenomenon. The 66 per cent of the radiation absorbed by the Earth is divided in the following ways: 23 per cent of its energy goes to evaporate water from oceans and lakes; 42 per cent goes towards heating the air, ground and sea; just 1 per cent goes to drive the wind and the waves on the sea's surface; and a mere 0.023 per cent is gathered by plants and other organisms to drive photosynthesis, converting carbon dioxide to oxygen and building up vegetable material at the bottom of the food chain, which higher life-forms can metabolize.

Photosynthesis is carried out by the complex molecule chlorophyll, of which there are several forms. All plants, and cyanobacteria, have chlorophyll a. It absorbs violet-blue and orange-red wavelengths from sunlight, and because it reflects the green, that is the colour of plants on Earth. Chlorophyll b and some other pigments in plants absorb more of the red and infrared wavelengths. These wavelengths correspond to the main radiation that reaches Earth's surface from the Sun. PhD student Jack O'Malley-James of the University of St Andrews (supervised by Jane Greaves, John Raven and Charles Cockell) assessed the potential for photosynthetic life in planetary systems orbiting different solar-type stars, for example red dwarfs. If plants exist on planets orbiting, say, red-dwarf stars, such as Gliese 581 (see p. 22), it is likely that they would have specialized to photosynthesize from the redder wavelengths. A red sun would not provide much blue or green light, so plants would adapt to absorb the red. On planets orbiting Gliese 581, black plants would blossom.

* * *

Although the Sun, to a large extent, is stable over long periods of time – a fact that has been benign to the evolution of life on our planet – it is variable on timescales that affect day-to-day life on Earth. The outer layers of the Sun are seething gas, circulating in vast convection clouds, tangling the Sun's magnetic field. Twists and breakages of the magnetic field tear apart the surface of the Sun, or cause it to spring in arches ('prominences') above the surface, into its atmosphere. These magnetic disturbances manifest themselves most obviously as spots on the Sun.

There are times when the Sun is very blemished: periods of so-called sunspot maximum, which recur every eleven years (21). This activity changes the radiation emitted, not so much the quantity but the spectral distribution: it will include more X-rays and ultraviolet light. This affects conditions in the upper atmosphere of the Earth. As a result, the weather changes and affects the growth of vegetation. The first time that this correlation was demonstrated was by the astronomer William Herschel (1738–1822) in 1801. He found that the wheat harvest and the price of wheat in England between 1630 and 1799 varied during the six sunspot cycles that he analysed; the more sunspots, the more abundant and cheaper the wheat. In the several centuries since, records have been kept about sunspot numbers and hundreds of studies have found similar correlations in cotton and many other crops, in North America and other regions.

The stars differ from one another in their number of spots, and therefore in their variability. The spots on the Sun – even at its most pimpled – cover no more than a few per cent of its surface. Some stars have spots that also come and go, but that at any one time may spread over an entire hemisphere. Their light output is erratic, and their magnetic variability is even more extreme. It is hard to see how planets orbiting such stars could be inhabited, since their natural ecosystem would be so disturbed.

The atmosphere of the Sun extends far beyond its surface (**22**) – indeed, all the planets of the Solar System orbit within it. Like that of the Earth, the solar atmosphere is variable, and produces 'space weather'. The most significant events of space weather are 'coronal mass ejections'. From time to time, but especially during sunspot maximum, there is a sneeze-like spasm on the surface of the Sun, and a cloud of solar material can be ejected into space, where it is denser than the Sun's typical atmosphere. The cloud may travel towards the Earth, which becomes bathed in an especially dense cloud of electrons, protons and other cosmic rays.

Electrically charged particles coming into the vicinity of Earth are diverted – mostly by the planetary magnetic field, acting as a shield – from the equatorial regions, and some spiral down the magnetic-field lines of the Earth into the north and south poles. When the particles strike the Earth's atmosphere they cause emissions from its gases: the aurora. This display happens about 100 km up in the atmosphere, at its top layers, and is considered an extremely beautiful sight by humans, looking up from the ground. The aurora is a sign that we are being protected from solar radiation; the layer of our atmosphere is equivalent to 10 tonnes per square metre of radiation shielding. The thin atmosphere of Mars means that it is more susceptible to the impact of solar cosmic rays; other planets and moons with no real atmosphere at all are even more vulnerable.

In space, of course, there is no atmosphere to protect from the damaging effects of radiation. Exposure can be risky to the health of astronauts or other living creatures. High radiation doses kill entire cells, while low doses tend to damage or alter the genetic code (DNA) beyond the built-in capacity of the cell to repair the damage. High doses can destroy so many cells that tissues and organs are damaged, with immediate fatal effects. The most dangerous time for an astronaut to be in space is thus during a coronal mass ejection.

Space agencies monitor the Sun for the earliest possible warning signs that a solar storm is threatening, in order to take evasive action. Any space walks are cancelled and the astronauts remain inside the protective skin of the spacecraft or space station. Spacesuits are made of a high-tech, multi-layered fabric blanket that maintains a life-saving atmosphere of oxygen and insulates against the frozen harshness of deep-space vacuum, but, at about 1 kg per square metre, this layer of clothing is not much more effective than gossamer against cosmic radiation. Safely inside their space vehicle, astronauts will reorient it so that its water tanks are upstream of the radiation, in order to give themselves maximum shielding (about 1 tonne per square metre). The cumulative radiation received by the astronauts is monitored so that they can be brought back to the ground if they approach their lifetime permitted dose.

There are some circumstances in which evasive action is limited and the exposure is prolonged. The trips by the Apollo spacecraft to the unprotected surface of the Moon lasted between one and two weeks, and were such cases. Astronauts received enough radiation to last a lifetime; they would not be licensed to work in the nuclear industry, as any further exposure to radioactive material would result in an overdose. The radiation that would be received in a longer manned spaceflight, such as the nine months it would take an astronaut travelling from Earth to Mars, is of major concern.

*   *   *

What of less advanced forms of life than human beings? Is everything living as vulnerable to cosmic radiation as we are? Is the idea that life might be transferred from one planet to another – the panspermia hypothesis, see chapter 7 – as dead as a mouse would be in a nuclear reactor?

Surveyor probes were among the first spacecraft to land on the Moon. Surveyor 3 landed in the Mare Cognitum portion of the Oceanus Procellarum in 1967. The same site was visited by the Lunar Module of the Apollo 12 manned lunar mission in 1969, and astronauts Pete Conrad (1930–1999) and Alan Bean (b. 1932) collected some parts of the Surveyor 3 lander, and brought them back to Earth, to see what the lunar environment had done to them. To everyone's surprise, fifty to a hundred bacteria were found deep inside the recovered camera, and were successfully revived. They were *Streptococcus mitis*, a common bacterium found in the mouths of humans, and which could have attached to the camera while it was being assembled. The bacteria would have survived two-and-a-half years of living in a vacuum, solar cosmic radiation and an average temperature of 20 degrees above absolute zero (–253°C), as well as a total lack of nutrients. In 1991, as Apollo 12's commander Pete Conrad was reviewing the transcripts of his conversations on the Moon, he noted in the margin that he had 'always thought the most significant thing that we ever found on the whole…Moon was that little bacteria who came back and lived and nobody ever said anything about it.'

Prompted by Conrad's comments, NASA reviewed this famous episode to be sure the bacteria had in fact been on the Moon, exposed to its malign environment, for all that time. The agency concluded that there had been too many opportunities for the bacteria to be added to the camera during its transit back to Earth in a duffel bag. Or, perhaps, the bacteria had never even been to the Moon, and were added during the camera's examination in the laboratory. It could have transferred after the camera was handled by laboratory equipment that had been laid on a bench contaminated by the sneeze of a member of technical staff (dignified in NASA-speak as 'a single non-sterile handling event'). These uncertainties mean that the discovery of live lunar bacteria is now regarded as insignificant.

In spite of this failed demonstration, dormant bacteria can be amazingly robust. The longest exposure time of bacteria to a vacuum was a bacillus that was revived after six years in a biological experiment. The well-controlled laboratory conditions proved that survival in a vacuum is quite possible. Some bacteria are also known to have endured in nuclear reactors. As long ago as 1958, operators at the Omega West reactor in Los Alamos noticed that the water in which its fuel rods were suspended, 8 m deep, was becoming cloudy. The water was a coolant and moderator, and also shielded the operators from the radiation emitted by the reactor. The cloudiness proved to be bacteria of the genus *Pseudomonas*. They were feeding on impurities that entered the water-circulation system, and had mutated to be able to survive nuclear radiation at doses that would kill a human being many times over.

Many other species of bacteria survive within nuclear reactors. Algae, fungi and yeasts have also been found in water covering the damaged core reactor at the Three Mile Island power plant in Pennsylvania, USA, and in nuclear-waste containers at Chernobyl. The most radiation-resistant bacterium (as certified by the *Guinness Book of Records!*) is *Deinococcus radiodurans*. Its name means 'strange berry that withstands radiation'. It was originally isolated from samples of canned meat that were believed to have been sterilized by a high dose of gamma radiation, but which subsequently went bad. It has been found in shielding pools for radioactive materials and is also able to live in dry, cold conditions, such as on Arctic rocks. Colonies of these bacteria could survive millions of years of exposure to the cold and radiation levels found in space or an airless planet.

The unusually high ability of *Deinococcus radiodurans* to repair itself from radiation damage suggests that it is unlikely to have evolved these characteristics through responding to any naturally radioactive environment here on Earth, because there is none extreme enough. It has been proposed that it may have developed on, say, Mars, and been brought to Earth on a meteorite. If that were true it would prove the viability of the panspermia hypothesis (chapter 7). From its DNA, however, it is clear that *Deinococcus radiodurans* is related on a phylogenetic tree to other terrestrial bacteria and must have developed here on Earth. It seems likely that the bacterium developed its remarkable abilities in response to some other stress, extreme cold or extreme dryness for example, and immunity to radiation damage was a by-product.

* * *

Coronal mass ejections from the Sun are highly damaging to life. Even if we are protected in some way from the direct impact of the solar cosmic radiation, it can still have a negative effect on the environment that supports us. Were these solar storms to impact with full force onto the Earth's atmosphere, they would gradually evaporate it. Our atmosphere survives because of the strength and persistence of the Earth's magnetic field, which deflects the brunt of the coronal mass ejections back into space, letting only a leakage impact at the north and south poles. The atmosphere is a vital protector of life on Earth also because it shields living tissue from the damaging intensity of solar ultraviolet radiation.

The Sun is the strong benevolent body that plays the greater part in maintaining our cosmic environment; other solar-type stars would do the same for some of the planets in their systems. But its sudden, violent outbursts are close to fatal and we are able to tolerate them only under cover of the Earth's atmosphere. The long-term increase in the brightness of the Sun will eventually produce the extinction of life on our planet, until in the end the Sun swells up into a red giant and, like the Titan Cronus, eats its children, swallowing the terrestrial planets and any life they might contain.

Over the past 4,600 million years, as the Sun has aged, it has (and will continue to) become brighter. This is because it is gradually using up its store of nuclear fuel, and the density and temperature in its interior regions is changing. When it first became a stable star its output was only about 70 per cent of the amount that it radiates today; we would expect that the average global temperature would have been much colder during the Archaean and the Proterozoic eons (see **Table 1**, p. 69), though the evidence is sparse and unclear. It is, however, clear that there were liquid water oceans on Earth by the end of the Hadean eon, so the greenhouse effect must have been more significant by that time, in accordance with the Gaia hypothesis (see chapter 8). The Sun will continue to brighten and eventually, perhaps in a billion years, the oceans will evaporate, the inevitable outcome put off only temporarily by the environment adapting as per the Gaia hypothesis. In any case, the Sun will run out of hydrogen fuel to such an extent that it will become a red giant, expanding out almost to the radius of the Earth. Our planet is likely to be absorbed back into the Sun, from which it originated. If for some reason it survives this fiery fate, it will travel with the Sun as, eventually, it fades away to become a white and then a black dwarf. Our planet will become as cold as if it had no Sun at all.

* * *

Some planets have met this fate of eternal cold already. In chapter 8 we saw how the chaotic interactions in a newly formed planetary system would eject some planets into space, breaking the gravitational bond with their star and becoming interstellar orphans. Ten such planets have been discovered by a new method that looks at the gravitational micro-lensing of stars in the Milky Way. There are surveys that regularly and repeatedly observe all the stars in an area of our Galaxy and measure their brightness. Some surveys like this are conducted to search for the minute dimming of a star caused by the passage across its face by a planet in orbit around it (see p. 21). Others look for a rare, but striking, phenomenon known as 'gravitational lensing', the sudden, otherwise inexplicable, brightening of a star. If a star or a planet, perhaps in orbit around a star or perhaps floating freely in interstellar space, passes almost directly on a line between us and a star that lies in the far background, the force of gravity from the mass of the foreground star or planet warps the space that surrounds it and acts like a magnifying glass. It is an effect of Einstein's general theory of relativity. The 'magnifying glass' gathers light from the background star and causes it to brighten considerably for a brief period of hours or days.

The lensing phenomenon was first found to occur when a foreground galaxy intensifies the light of a background galaxy, so it had originally been thought to take place in nature when the mass of the lensing object was very large. When a star or planet – much less massive than a galaxy – does something similar, it is called micro-lensing. Micro-lensing is rare because the alignment of the foreground star or planet and the background star has to be very good (though it does not have to be perfect), and of course the foreground star or planet continues on its wandering journey, so the alignment happens only once and you get only one chance to see the phenomenon. But stars are numerous, so many hundreds of micro-lensing events have been recorded. And some of them have been proven to be caused by planets.

A survey for micro-lensing called MOA (Micro-lensing Observations in Astrophysics) is based on a modest telescope at Mount John University Observatory in New Zealand, supported by OGLE (Optical Gravitational Lensing Experiment), a Polish astronomical project using telescopes at the Las Campanas Observatory in Chile. The projects have seen many events caused by a star passing in the foreground. They have observed a few cases in which a succession of micro-lensing events were visible, caused by a star and one or more planets in orbit around it. They have seen ten isolated micro-lensing events that indicate that the foreground object had the same mass as Jupiter, showing that it was a planet rather than a faint star. The ten micro-lensing events took place with no other foreground star involved.

There is the possibility that some of the ten planets may move in very wide orbits around stars, so a foreground star could have been involved, but a long way away from the line of alignment. But it is believed that nature almost never makes jupiters in such distant orbits. It is more likely that these would be dark, isolated planets, free-floating in space, far from any parent star. Extrapolating from the ten cases, the survey suggests not only that these planets exist, but also that they are numerous: there are twice as many free-floating jupiters as stars. It seems likely that there are at least as many free-floating smaller planets of a similar size to Earth.

We might expect a free-floating earth to be frigid, literally frozen at such a distance from the warmth of any stars. While the Earth receives most of its heat from the Sun, however, it does also produce its own from the decay of radioactive elements within its core, and is still using some of the heat from its pounding by planetesimals during its formation (these three sources of heat are mentioned in **Table 5**, p. 124). Planetologist David J. Stevenson (b. 1948) at the California Institute of Technology and geo- and astrophysicists Dorian Abbot (b. 1982) and Eric Switzer (b. 1981) at the University of Chicago have investigated the habitability of free-floating Earth-like planets with similar mass, water content and composition of radioactive elements, and atmospheres of hydrogen and helium, or carbon dioxide. According to Abbot and Switzer, planets that were somewhat larger than the Earth and with the same water content, or planets the same mass as the Earth with substantially more water content, could support a liquid water ocean under a layer of frozen ice and a frozen carbon dioxide atmosphere, somewhat like Jupiter's moon, Europa (chapter 14). Abbot and Switzer refer to free-floating earths with oceans under an ice cover as 'steppenwolf planets' because 'any life in this strange habitat would be a lone wolf, wandering the galactic steppe.'[1] Of course, it is one thing to say that such a lone planet with liquid water could exist, and another thing to prove that it does – let alone that there is life in those oceans.

In the future, telescopes that are sensitive to the infrared might detect steppenwolf planets. But the detection of life on such an object, should there actually be any, is very far in the future indeed.

11

Life and the Climate

Earth has a climate conducive to life, but even on a planet with such a benign environment, extinctions are extremely common: the vast majority of species that have ever lived on Earth are now extinct. Are we lucky to have survived so long, and is it likely that life-forms like us on other planets could have died out before we have even had the chance to discover them?

The worldwide variation of climate is caused fundamentally by changes in the amount of solar radiation striking the surface of the Earth, as well as by the ways in which this heat interacts with the atmosphere. Our planet is a spinning sphere, with the axis of spin approximately at right angles to the direction of the Sun. As the Earth spins, any given place on its surface will face alternately towards and away from the Sun; this causes the daily cycle of light and heat. When sunlight directly illuminates the surface of the Earth, it has a more concentrated effect than when it strikes the ground obliquely; sunlight and heat are more intense when the Sun is directly overhead. This means, of course, that it is warmer in the middle of the day and at the equator than as the Sun rises and at the poles. Life is usually more active and abundant in the presence of solar radiation, up to a point, because such radiation is a form of energy that can be utilized directly, and because the biochemical reactions that energize life will, in general, happen more quickly when it is warm.

Organisms adapt to their location by developing defensive strategies against extremes of cold and heat. In evolution, one simple stratagem to avoid heat loss in cold environments is to insulate, for example by means of feathers or hair. Another is to alter size. The amount of heat that a living body generates depends on the amount of biomass in the body, which in turn depends on its volume and therefore the cube of its dimensions. The amount of heat that the body will lose relies on its surface area, or the square of its dimensions. Of two similarly shaped life-forms in the same environment, the larger one will be the warmer because it generates proportionately more heat than it loses. It will therefore be more suited to a cold climate. Conversely, in a hot climate the life-form would need to dissipate heat more quickly and a high ratio of surface area to volume would be beneficial, so being smaller would be more efficient. This general tendency goes by the name of Bergmann's rule, formulated by the German biologist Karl Bergmann (1814–1865), in 1847.

Other strategies to cope with cold are to prevent freezing of fluids in body cells by chemical means. When body fluids freeze they are less able – perhaps unable – to perform their biological function. Moreover the ice crystals that form are larger than the equivalent amount of water and so freezing causes physical damage to body tissues, bursting apart the cell walls and organs. An organism can avoid this by replacing its internal water with a fluid that freezes at a lower temperature. Tardigrades, or 'water bears', are microscopic aquatic creatures, perhaps up to 1 mm in length, which have eight legs and appear to waddle like a bear (hence their common name). They can survive freezing temperatures by replacing their internal water with the sugar trehalose. Tardigrades can also survive dehydration by changing the glucose in their bodies to trehalose, a process that reverses when they are revived by soaking in water. In fact some tardigrade species are able to withstand a range of extreme environments: both very low and very high temperatures, extreme intensities of nuclear radiation, and even years without water. For ten days in 2007, tardigrades survived exposure to cosmic radiation, vacuum and low temperatures in space on the Russian-ESA space mission, FOTON-M3.

Both simpler and more complex creatures survive freezing temperatures through the use of AFPs (antifreeze proteins). These compounds coat the surfaces of ice crystals and prevent them growing further – the difference between this mechanism and that of a car engine's antifreeze (in which ethylene glycol dilutes the water and lowers its freezing point) has led to moves to rename AFPs more appropriately, to avoid confusion. AFPs are found in some insects, plants, fungi, fish and diatoms. They are found in bacteria from lakes in Antarctica that contain microbial communities, but no fish and few zooplankton species. Despite the harsh environment, the communities in these lakes function, even during winter, thanks to AFPs.

* * *

The reason for the change of climate on a yearly cycle – what we ordinarily call the change of the seasons – is that the spin axis of the Earth is not exactly at right angles to its orbital plane and the direction to the Sun: there is a misalignment called the 'obliquity angle' of the Earth's polar axis, which is tilted by approximately 23.4°. Over short timescales, a year or a century, the axis of spin maintains its direction in space while the Earth is carried in its orbit around the Sun, so one pole of the Earth is tilted towards the Sun when the planet is on one side of its orbit and, six months later, the same pole is tilted away. The warmest zone on the Earth is in general the equator, but more precisely the sub-solar latitude. The sub-solar latitude changes through the year, being 23.5°N (the Tropic of Cancer) in June and 23.5°S (the Tropic of Capricorn) in December. This produces the annual cycle of the seasons. The northern hemisphere is tilted towards the Sun during its summer, and away from the Sun during winter. The South Pole does the same, but six months out of phase, and the southern-hemisphere summer occurs when the north is experiencing winter, and vice versa.

The tilt of the Earth by this large amount, 23.4°, is thought to have originated with the impact of Theia with the Earth, when the Moon was formed. Evidently the impact was not square on, and the Earth's axis was knocked askew. The cycle of the seasons – the poetry of spring, the harsh grip of winter, the torrential rain of the monsoons, the drying heat of the sirocco wind – can be traced back to this chance event.

A smaller but nevertheless noticeable effect on the annual cycle of the weather is the eccentricity of the Earth's orbit around the Sun, such that the Earth is closest to the Sun in the first week of January and furthest from it in the first week of July, the difference amounting to 3.4 per cent, and causing a 7 per cent change in the amount of solar radiation. More radiation strikes the southern hemisphere during its summer than occurs in the northern hemisphere's summer.

Over long periods of time (tens of millennia) the Earth's axis does not always point in the same direction, and the eccentricity of the Earth's orbit changes. The Earth's axis wobbles, like the axis of a spinning top, so that it maps out the surface of a cone (the technical term is 'precesses') over a period of 26,000 years, so on that timescale the seasons will shift. Moreover, the tilt does not remain constant at 23.4°, but varies between 21.5° and 24.5° over a period of 41,000 years. The larger the tilt, the greater the variation of the seasons. The range is significant, but in reality the change is rather limited. The reason for this is that the gravitational pull of the Moon stabilizes the wobble of the Earth, holding its rotation axis steadier than it would have been without the Moon being close by and so large: another favourable outcome of the Big Splash (see chapter 8).

As well as this wobble in the Earth's axis, the eccentricity of its orbit is also variable and changes between almost nothing and twice its current value on a timescale of 100,000 years. It is thought that these orbital fluctuations have been responsible for the recent natural changes in climate, including the coming and going of the Ice Ages. An 'ice age' is a period when the temperature of the Earth's surface and atmosphere is so much reduced that there are large areas of ice on continents in both hemispheres, ice sheets at the poles and glaciers slipping down from mountains. By this definition, the existence of the Greenland and Antarctic ice sheets currently places us in an ice age. What is referred to as the Ice Age is the most recent ice age, during which large parts of America and Europe were covered with ice, with glaciers penetrating down to the latitude of what are now the northern states of the USA and the northern halves of Europe and Russia. The first evidence for the existence of the Ice Age was the geological observation that the glaciers of the Alps were once more extensive than they are now: rocks were scoured by scratches produced by glaciers ('glacial striations'), glacial deposits of rock were more widely distributed, and so on. More direct evidence of the way the temperature of the Earth has varied comes from measurements of isotopic ratios in cores of ice drilled up from undisturbed ice layers from the Antarctic and from Greenland, and from ocean sediments.

The longest record of the temperature over the last 420,000 years has been measured from the deepest ice core yet drilled: 3,623 m deep, at the Russian Vostok station at the centre of Antarctica. The largest archive of ice cores is held at the US National Ice Core Laboratory in Denver, Colorado. It has more than 14,000 m of ice cores from 34 drill sites across Greenland, Antarctica, and high mountain glaciers in the western United States. The facility is maintained at a temperature of $-35°C$, with four levels of backups and safety systems, to preserve the cores for study. Within each core the ice is layered because snow fell more at the same season of an annual cycle; the dates can be established by counting the ice layers down from the top. Air bubbles in the layers also provide information about the atmospheric composition through time.

A water molecule is made up of one oxygen atom in combination with two hydrogen atoms. The oxygen atom exists in two stable isotopes, ^{16}O and ^{18}O, and water molecules can be lighter or heavier accordingly.[1] When water evaporates from the sea or from a lake, the lighter water evaporates first, so water vapour in the air has more ^{16}O than normal. But the bias in the ^{16}O depends on the temperature; the hotter it is, the easier it is to evaporate ^{18}O. When the water vapour condenses and falls as snow onto an ice sheet, the isotopic ratio preserves a record of the temperature of the evaporation, linked to the global temperature. There are similar changes in isotopic ratios in oceanic sediments and in rocks that can be dated by radioactive means, but these give a less densely sampled and perhaps less accurate temperature series.

The available evidence suggests that the present ice age began 40 million years ago. It grew colder during the Pliocene and Pleistocene periods, starting around 3 million years ago, with the spread of ice sheets across the northern hemisphere. Since then, there have been cycles in which glaciers have advanced and retreated every 40,000 to 100,000 years. The most recent retreat of the glaciers ended about 10,000 years ago.

The Serbian civil engineer and geophysicist Milutin Milanković (1879–1958) devoted his later career as a mathematics teacher in the 1920s–40s to correlating the effect of cyclic variations of the Earth's orbit with the global temperature. The cycles in the orbital parameters are usually referred to as Milankovitch cycles in countries outside Serbia, because the pronunciation of the final character of his name is unfamiliar. Milanković himself attributed the Ice Ages to these effects. In particular he discovered the main recent periodicity of about 100,000 years (the result of a combination of orbital effects). The same periodicity is discovered in the record of global temperature preserved by the ice cores and the ocean sediments.

* * *

Milankovitch cycles are not large enough, on their own, to alter global temperature by the amounts that are recorded, and there must be some further effects that amplify the repercussions of changes in solar radiation. The atmosphere of the Earth has a big influence on the distribution of warmth on our planet, because of the greenhouse effect – discovered in 1827 by Joseph Fourier (1768–1830), the French mathematician. Sunlight and solar infrared radiation warm the land, atmosphere and oceans. These warm masses radiate back towards space in infrared wavelengths. But the infrared emitted from the surface is mostly absorbed in the atmosphere by greenhouse gases and clouds. Infrared radiation thus does not escape into space, but warms the lower air and keeps the surface hotter than otherwise expected. On Earth, at the equator, especially over Africa, the temperature is 25 per cent warmer due to the greenhouse effect (10 per cent is equivalent to 31°C). Over North America, Australia and most of Asia the temperature is 10 per cent warmer. Even in the polar regions of the Earth, the temperature is a few per cent warmer because of the greenhouse effect. A reduction of just 10°C in the average temperature is sufficient to plunge the surface of the Earth into a major glacial period, in which ice sheets could extend from the poles to the equator.

Water vapour, carbon dioxide and methane are the most important greenhouse gases on Earth at the present time. The Swedish chemist Svante Arrhenius speculated that it was changes in the levels of carbon dioxide in the

Earth's atmosphere that altered its surface temperature through the green-house effect. In the last century or two, the burning of fossil fuels (coal and oil) has increased the amount of carbon dioxide. The greater areas of land used to maintain animal herds has also led to an increase in the amount of methane, which is emitted by flatulent agricultural animals, particularly cows and other ruminants. At the same time, deforestation has both released carbon dioxide into the atmosphere through the burning or decomposition of trees, and reduced the density of leaves that withdraw carbon dioxide from the atmosphere into the biomass. These increases in greenhouse gases are responsible for the anthropogenic component of global warming.

The greenhouse effect is not unique to planet Earth. As long ago as 1908, astronomers were aware of the same phenomenon in the atmospheres of other planets, including Venus.

The greenhouse effect has changed through geological time, because the composition of the atmosphere has shifted such a lot. In some eras there were more active volcanoes than there are today, emitting copious quantities of carbon dioxide, methane and water vapour, which increased the greenhouse effect. There have also been systematic changes associated with the evolution of life on Earth, in ways that demonstrate the Gaia hypothesis (see chapter 8). Early on in the Hadean and Archaean eons there was far more carbon dioxide and methane in the Earth's atmosphere than there is now, before much of it was fixed in geological deposits and in the biomass of vegetation, and the carbon dioxide of the atmosphere was replaced by oxygen. This warmed the Earth even though the Sun was cooler. We know that oceans existed on Earth at that time, as copious amounts of liquid water rather than ice. At the end of the Archaean eon and the time of the Great Oxygenation Event, when there was a surge in the amount of algal vegetation, with carbon dioxide and methane becoming fixed in the biomass, the greenhouse effect was much reduced and the Earth cooled in what is known as the Huronian glaciation, which extended from 2,400 to 2,100 million years ago. There were further major glaciations 730 million years ago, and about 560 to 650 million years ago, with the last glaciation finishing just before the Cambrian explosion, for which it may have been a trigger.

Some scientists identify these great glacial periods, during which the entire Earth may have been covered in ice and glaciers, as 'Snowball Earth' events. (Whether or to what degree they happened and, if they did, what caused them, are controversial topics.) The main evidence is that there are vast deposits of rock originating under glacial conditions in the geology of rocks of this age in some areas near the equator, far from the cold, polar regions. The cause of Snowball Earth is proposed to be that once the Earth cools for some reason and a glaciation starts, the highly reflective ice cover increases the amount of solar radiation that is reflected back into space and the Earth cools further, progressively extending the reach of the glaciers. The difficult part seems not to be the making of Snowball Earth, but rather its undoing, recovery and the end of glaciation. Perhaps the end of Snowball Earth is the continued eruption of volcanoes and an increased greenhouse effect, because with ice covering the land and sea, carbon dioxide cannot be absorbed from the atmosphere, so its concentration builds up. Perhaps Snowball Earth brings about its own demise.

The first Snowball Earth episode took place in association with the Palaeoproterozoic Great Oxygenation Event, in which the surface of the entire planet was frozen from poles to equator, from about 2,300 to 1,800 million years ago.

The second Snowball Earth happened around 730 million years ago, when the weathering of silicate rocks and the consequent increase in levels of oxygen affected the climate. Reductions in the methane and carbon dioxide levels reduced the greenhouse effect, the temperature fell and the Earth began to freeze, in what is known as the Sturtian ice age. Again, the planet was completely covered with ice and glaciers until about 670 million years ago.

The third extreme glaciation of the Earth occurred between 640 and 580 million years ago. The glaciation is known as the 'Marinoan' and tailed off into the 'Gaskiers' glaciation. These ice ages were probably a result of changes in the greenhouse gas content of the atmosphere, caused by the Great Oxygenation and by massive volcanic eruptions.

Snowball Earth events must have brought about major changes in the evolution of life on our planet, probably mass extinctions. In a mass extinction, numerous species die off at the same time and from the same cause. It is estimated that, of the 5 billion to 50 billion species that have existed on this planet, just 50 million remain alive today: as much as 99.9 per cent of all species that have ever lived are now extinct. The precursors of our own species – all the Australopithecines, *Homo neanderthalensis*, *Homo erectus*, and *Homo habilis*, to pick only the best-known examples – have all disappeared, and, eventually, this will happen to us, *Homo sapiens*, too. We have become a very capable species, and although we might postpone the inevitable by altering our environment instead of ourselves, nature will win in the end, either in a dramatic disaster (described in the next chapter) or in a less cataclysmic, yet still irreversible, way. According to the geologist Gerta Keller (b. 1945) of Princeton University, who is well known for her theory on the extinction of the dinosaurs (see chapter 12), 'Of course we, too, will be wiped out, eventually. Something else will evolve, perhaps similar to us.... The dinosaurs, after all, turned into birds. We will be a footnote to others who will ask themselves why we died out.'

Extinction has thus been the norm, but usually one species at a time. Mass extinctions are more dramatic but account for less than 5 per cent of the total. The nature of the extinctions and changes in the evolution of life during the Snowball Earth events is less than clear, because the evidence is so fragmentary. We could expect that the freezing temperatures would have slowed down the metabolism of the creatures that were alive at the time, the low levels of methane reduced the intake of nutrient gas on which they relied, and the presence of increased oxygen interfered with their metabolism. This would have killed off numbers of the prokaryotes and eukaryotes that were not able to adapt to the environmental changes that took place.

There is some indirect evidence about the quantity of biological activity during the ice ages. There are two stable isotopes of carbon in minerals dissolved in seawater: the most common, at 99 per cent, is carbon-12 (^{12}C) but there is 1 per cent of carbon-13 (^{13}C). For reasons explained on p. 121, the process of photosynthesis tends to fix the lighter isotope from carbon dioxide into the biomass of the photosynthesizers. Bacteria and algae thus tend to have less ^{13}C than the air and the ocean water in which they live, relative to ^{12}C. Sediments deposited from life-forms preserve the depletion of ^{13}C. There are well-defined strata from the Neoproterozoic era (1,000–542 million years ago) in the Mackenzie Mountains in north-west Canada, Spitsbergen in Norway, and Namibia, which show a number of periods when there were strong reductions, although it is not clear exactly when these were, and whether they occurred at the same time everywhere.

Although the glaciations may have been large-scale, there would have remained niche environments where more equable temperatures were maintained – for example, pockets of warmth in the oceans near areas of volcanic activity, such as 'black smokers' (see chapter 7). And, as we have seen, organisms have an amazing capacity to adjust and adapt to extreme environments. The success of terrestrial life in dispersing itself over the whole world in response to the challenges posed by variations in climate has had a beneficial effect on biological diversity and its evolutionary progression. This suggests that extraterrestrial life will be able to respond well to the climatic variations of its world, provided that such a world is basically habitable. This in turn gives cause for optimism that we will find life on some of the planets in our Solar System with challenging climates, such as Mars.

We can also anticipate that extraterrestrial life on worlds with stimulating weather will have developed in response to that environmental stress, and one outcome of this evolution could well be intelligence. We may find, if and when we meet a humanoid extraterrestrial alien, that we both have the same interest in the weather as a subject for small talk, before getting on to the serious conversations about interstellar propulsion systems and fostering world peace.

12

Considering Disasters

Life exists on our planet, and presumably others, as part of the cosmic environment – we are *a part of* the Universe, not *apart from* it. As such, life on Earth is subject to changes in its environment not only on a regional or global scale, but also on a cosmic one. As I have discussed in earlier chapters of this book, such changes can be gradual and evolutionary, as in the long-term variations in the Sun, but also major and sudden. We can usefully go back to the origin of the word 'disaster' to label such changes. One of the Latin words for 'star' is *astrum*; together with the prefix *dis-*, meaning 'away', the word *disaster* embodies what will happen should there be a negative influence from space. Another Latin word for star is *sidus*. In order to know the worst, a person might inspect the night sky and think about what the stars foretell – from *con*, 'together', and *sidus* comes the word 'consider' – which is what will happen in this chapter, and in this book.

The effect of large-scale natural disasters on the ecosystem can be very extensive, leading to such environmental stress that many species are unable to survive, although some others will flourish. This stress can take the form of a sudden and profound change in the global climate caused by a cosmic catastrophe. Such shifts are sometimes the cause of the discontinuities that separate the geological eras and epochs. Even if the effect of such a catastrophe is more local, the consequences for some of the species abundant in that region can be devastating, and their extinction can proliferate beyond the range of the catastrophe as new species move in and compete for the niches they used to occupy.

* * *

Probably the best-known proposed cause of a mass extinction on Earth is an asteroid or comet impact. 'Everybody knows' that a meteor impact killed the dinosaurs – although the truth is probably more complex than this, as we shall see. Meteor impacts still happen. The episodes of the Late Heavy Bombardment and the creation of the Kuiper Belt and Oort Cloud cleared the Solar System of large numbers of asteroids and comets, and the major planets swept clean their feeding zones of nearby planetesimals. This reduced the potential for meteor impacts on Earth. Nevertheless, some asteroids and comets remain in the Solar System. Mostly they orbit safely in the Main Asteroid Belt, the Kuiper Belt or the Oort Cloud. They can be nudged from their orbit, however, by interacting with one another and by the random gravitational pulls of the major planets. The new orbit may bring the asteroid or comet across the path of Earth, and with it the chance of a collision.

The hazard associated with an impact depends on the size of the meteorite, from what material it is made, and where on Earth it strikes. Small meteoroids (up to a metre in diameter) are numerous and frequently collide with Earth. They produce nothing more than a meteor or fireball in the sky, and the meteoroid is usually completely disintegrated by the heat of its passage through the air into dust. If a meteoroid up to about 10 m in diameter strikes the Earth, its surface likewise disintegrates and falls off during its bright passage through the air. It may crack into several pieces, and a residual rock may strike the Earth as a meteorite, producing a shallow hole in the ground. A mid-sized asteroid (say 500 m in diameter) will cause an atmospheric blast wave, earthquakes, a big hole in the ground and considerable heat radiation from a fireball if it crashes to land. Since two-thirds of the world is covered by water, however, it is more likely to fall in the ocean, where the impact will probably generate big waves that propagate over great distances and surge onto the shore, causing floods, depositing sediment, and washing out coastal plains. At the present time, when such a substantial part of the human population lives near the shore, the risk posed by these impact-generated waves is high, although the asteroid blows that cause them are rare. The tsunamis of December 2004 in the Andaman Sea and of April 2011 in Japan (caused not by asteroid impacts but by undersea earthquakes) have shown the devastating effects of waves of only moderate height, compared to other events on the geological timescale. The tsunamis have a direct effect not only on people but also on anything built near the shoreline, such as homes and businesses, or a nuclear power station.

The greatest risk to life on Earth is from large-diameter asteroids: those greater than a kilometre in diameter. The father-and-son team, Nobel Prize-winning physicist Luis Alvarez (1911–1988) and geologist Walter Alvarez (b. 1940), both from the University of California at Berkeley, studied the effects of the impact of large asteroids on Earth: they do nothing less than completely alter the global climate. In a paper in the journal *Science* in 1980, 'Extraterrestrial cause for the cretaceous-tertiary extinction', which caused a sensation and which has been cited in the scientific literature more than 400 times (an extraordinarily high number), they wrote:

> The impact of a large earth-crossing asteroid would inject about sixty times the object's mass into the atmosphere as pulverized rock; a fraction of this dust would stay in the stratosphere for several years and be distributed worldwide. The resulting darkness would suppress photosynthesis, and the expected biological consequences match quite closely the extinctions observed in the paleontological record.

The Alvarezes were referring to several mass extinctions in the geological record that they believed might have been caused by asteroid impacts. The best-established – but as we will see, not the unequivocal – cause of extinction is the one they identified in 1980 as due to a meteor impact: the Cretaceous-Tertiary extinction that took place as the Cretaceous period was replaced by the Tertiary.

*   *   *

The Cretaceous period began around 145 million years ago and ended suddenly 80 million years after that. Before then, reptiles, including dinosaurs, ruled the land; afterwards, many of them – also including the dinosaurs – became extinct, and the age of the mammals began. The end of the Cretaceous period is marked by abrupt changes in the rocks laid down in the geological record, the so-called Cretaceous-Tertiary boundary (abbreviated as the 'K-T boundary,' from its German initials).

Since about 1970, Walter Alvarez had been studying the Bottacione Gorge in Gubbio, near Assisi, Italy, intending to correlate the age and magnetic properties of the 400-metre-thick limestone rock from the Cretaceous period that had been conveniently exposed by the roadside. The limestone is pink (the rock's name in Italian is *scaglia rossa*, meaning 'red flakes'), indicating that it contains concentrations of oxidized iron, similar to rust. This makes the rock particularly suitable for studying the history of the changes in terrestrial magnetism, stratum by stratum, dating each layer by the fossils that it contains. In 1977, Alvarez visited the place near the top of the limestone where the K-T boundary could be dated by the sudden absence of the species *Foraminifera globotruncana*. The foraminifera are small animals that live on the bottom of the sea, an environment ideal for their future fossilization. They leave distinctively shaped, shell-like fossils. They have been present in many different varieties since the Cambrian explosion 542 million years ago; their existence proves very useful for dating rocks.

The K-T boundary Alvarez found in the roadside cutting was marked by a thin (about 1 cm thick) layer of clay, sandwiched by the limestone strata (**23**). Something similar is present at the K-T boundary embedded in rocks in many different parts of the world. Clay is made of pulverized, flour-like rock-dust and water, and the layer spoke of the events associated with the changes and extinctions – although what it said was not entirely clear. In its broadest terms, the clay represents the change of the environment 65 million years ago that caused the mass extinction. Alvarez took a sample of a small piece of the rock containing the embedded clay layer and brought it home to examine more closely. According to the particle physicist Charles Wohl, writing about Luis Alvarez, Walter happened to show the sample to his father, telling him that the layer marked where the dinosaurs and much else went extinct, but that nobody knew why, or what the clay was about. This intrigued the physicist Luis, although he had never shown much interest in his son's work in geology before.

Looking for clues as to how the clay stratum had been laid down, Luis Alvarez decided to measure the concentration in the layer of iridium, a metallic element that is very rare on the surface of the Earth – indeed, it is believed to be the rarest terrestrial element. The reason for this scarcity dates back to when Earth was hot and its iron melted, sinking deep into the planet's core. Iridium – like other metals termed 'siderophiles', or 'iron lovers' – has an affinity with iron, and at this time it followed it down into the fiery depths. The elements that made up the Earth thus separated. The siderophiles, including iridium, are abundant deep in the Earth's core, but the surface material of our planet is greatly depleted of them.

Iridium is distributed throughout the Solar System, and appears in the same abundance as it does in the Sun in such material as that of small meteorites, the elements of which have not separated, because they have never melted. In other, larger asteroids, iridium sank into their iron cores, as happened to the Earth; but, when some of these asteroids broke up in collisions, their cores fragmented and became meteoroids, with concentrations of iridium in conjunction with the iron. Meteoroids of both sorts thus contain much greater amounts of iridium than the surface of the Earth. When these meteoroids encounter our planet and burn up in its atmosphere, iridium is thus deposited as meteor dust. The amount of dust is small, but over time it accumulates on the surface of Earth – part of the soil in your back garden has been deposited there in this way.

The background rate at which meteors strike the Earth is approximately constant, although there are some additional, spectacular, meteor showers – some annual, some sporadic. But, by and large, the amount of iridium in a rock measures how long it took to be deposited there. If the clay was laid down slowly, over a long period of time, that would indicate one kind of change from the Cretaceous to the Tertiary periods, but if the clay was laid down in a considerable quantity in a short period of time, as in a sudden flood, then that would indicate another. Luis Alvarez worked with nuclear chemists at the Lawrence Berkeley Laboratory, a renowned US government-funded laboratory of the Department of Energy located near the University of California's Berkeley campus, to measure the iridium in the clay. They used a technique in which the material is bombarded with neutrons to make it radioactive, with the amount and kind of radioactivity generated by iridium being easy to measure.

The result was unexpected: in the clay, at the K-T boundary, the iridium was a hundred times more abundant than is typical on the surface of the Earth. The chemists found the same phenomenon at another clay layer at the K-T boundary, this time exposed in a cliff in Denmark. In fact, an iridium layer turned out to be a common feature at the K-T boundary. This super-abundance of iridium was not the result of the steady fall of meteorite dust. Something much bigger must have caused this anomaly. It gradually dawned on the Alvarezes that the iridium layer indicated that a single, large meteorite had fallen to Earth at that time, disintegrating in the impact and spreading iridium over the world – or at least over a significant fraction of it. The body had been about 10 km in size – an asteroid rather than a meteoroid.

It was not long before about a hundred clay layers at the K-T boundary all over the world yielded other signs of a meteorite impact. They contained tiny, glassy beads and diamonds, created by the heat of the impact on the ground, and fragments of quartz that had been shocked by its enormous pressure, their crystal structure distorted under the strain.

Alvarez calculated the effects of such an impact. There would be a substantial crater excavated at the impact site, and enormous amounts of dust – hundreds of thousands of cubic kilometres of it, say a million billion tonnes – would be injected into the atmosphere, stopping sunlight from reaching the ground, creating a long winter, lasting many years, until the dust settled. The scenario was similar to that calculated for a global nuclear war – in fact, the concept of a 'nuclear winter' was put forward in an article 'The Atmosphere after a Nuclear War: Twilight at Noon', by the Dutch scientist

Paul Jozef Crutzen (b. 1933) and the Colorado-based chemist John Birks (b. 1946), in 1982, two years after Alvarez's work on the effects of a meteorite impact.

The reality of the meteor strike was confirmed in 1990 by the identification of the crater caused by the impact. When a meteor strikes the Moon's surface, it will certainly make a crater, the traces of which – a big hole, the rampart mountains that surround it – will survive for ever, unless a later meteor hits the same spot, because there is little on the Moon to fill up the hole, or wear the mountains down. When a meteor strikes the Earth, however, it may plunge into the sea, and cause no crater; it may make a crater that is subsequently covered by the sea and filled in by silt; it may also create mountains that are worn down by wind and rain, the residual sand from which is blown by the wind into the crater hole. The weather levels the ground and conceals the huge disaster that the meteor has caused. So, the only impact traces that are readily visible on Earth are those made on solid ground, and recently.

Nevertheless, after the surface signs of an old crater have been worn away by erosion, but before it disappears completely, subtle traces may remain, concealed under the weathered surface. This turned out to be the case for the K-T impact crater. In 1978, geologists exploring for oil in the Yucatan Peninsula in Mexico, near the fishing village of Chicxulub, found the residual traces of a huge crater, 200 km or more across. Later discoveries suggest that it is 300 km in diameter. The initial work was not released, as a matter of commercial confidentiality. The crater was first publicly recognized in 1990 by Canadian geologist Alan Russell Hildebrand (b. 1955) of the Geological Survey in Canada during work on his doctoral dissertation. When mapping large areas of ground, geologists use the technique of 'gravity mapping' to look below the surface. The varied density of the rocks below causes the force of gravity to be different above. A map of the force of gravity reveals the structures that lie beneath the ground (24). A hole in a rock that has been filled in by drifting sand is less dense than its surroundings. By contrast, if the hole has been made by an asteroid impact, the roots of the rampart mountains of the crater will have been compressed, and are therefore denser than the filled-in hole. The geologists found such a structure in the gravity map, in rocks of the right age to coincide with the K-T boundary. It was a hole at least 170 km in diameter, overlapping the Mexican coastline. Here, it is believed, is the meteor crater left by the impact associated with the extinction of the dinosaurs at the end of the Cretaceous period.

The location of the crater suggested that the impact would have caused a giant tsunami spreading across the Gulf of Mexico into the Atlantic Ocean, as well as a huge hole and tonnes of atmospheric dust. It is estimated that up to 85 per cent of all species were destroyed in the K-T extinction event, including plants, marine reptiles, diatoms and nanoplankton – as well as the land-dwelling dinosaurs. Surprisingly, a seemingly arbitrary range of species was spared, such as amphibians, flying dinosaurs (birds), crocodiles, ferns, insects, lizards, seed-producing plants, snakes and turtles. The most important survival, from a purely selfish human point of view, was the sparse group of small mammals that coexisted with the dinosaurs, which then diversified and became dominant on land. By exterminating the dinosaurs, the K-T event created room for us. Before the K-T event, dinosaurs prevailed; since then, it has been the age of the mammals.

* * *

Not everyone was convinced by the explanation arrived at by Luis Alvarez and his son Walter. The key question is not whether an asteroid struck our planet and created the Chicxulub crater and an iridium layer of dust across the world – the evidence for that is convincing – but rather, did this event coincide with, even cause, the global change that resulted in the K-T extinction and affected, later, how rocks were deposited? Gerta Keller, a geologist from Princeton University, has led the argument that it did not, claiming instead that the impact predated the K-T extinction by 300,000 years.

Keller studied rocks of the relevant age that had been laid down in Texas, in circumstances that seem not to have been disturbed by the impact itself, and have remained undisturbed by subsequent upheavals – earthquakes or volcanoes – in the surface of the Earth. The strata preserved in detail the sequence by which the rocks were formed, their composition, and the fossils deposited in them. Keller found that there was indeed a 3 cm layer of clay in the strata that contained glass beads flung out in a molten shower by the Chicxulub impact. It was below a layer of sandstone that had previously been hypothesized to have been deposited by the tsunami impact, and above which was the K-T boundary. According to Keller's interpretation, the clay had been deposited on the floor of a shallow sea, when the impact occurred and laced the clay with the glass beads. Over tens of thousands of years afterwards, the sea level fell and sand from eroding shoreland was deposited on top of the clay. Then the climate altered and dramatically changed the deposition of rock, effecting the transition from the Cretaceous period to the Tertiary.

So Keller concluded that the K-T extinction event took place long after the Chicxulub meteor impact. If the meteor was not the cause of the extinction, what was? Keller links the extinction with massive volcanic activity that was taking place at that time in India, creating the world's largest volcanic plain, the Deccan Traps. The lava there is 2 km thick, and originally covered a million square kilometres. The volcanic eruptions that created it lasted about 40,000 years, outgassing sulphur dioxide, carbon dioxide and hydrochloric acid. These gases would have caused global acid rain. The volcanoes also created drifting clouds of millions of tonnes of volcanic ash that blocked sunlight from reaching the planet's surface, caused the temperature to fall and altered the chemistry of the Earth's oceans and atmosphere. Here, Keller believes, is the cause of the climate change and the consequent alteration of geological processes and the evolution of life: they do not derive from the impact of an asteroid.

Thus Keller's image of the K-T extinction differs from Alvarez's. What we know for sure is that the K-T extinction occurred at about the same time as the meteor impact and the volcanic eruptions. Whether one or the other was the only cause has been a topic of heated controversy. In an attempt to reconcile the opposing views, some geologists have supposed that it may have been a coincidence of disasters, including asteroid impact and volcano activity, which triggered the K-T extinction event. The dinosaurs suffered from a succession of stresses caused by at least two geological events, and perhaps more, and were forced to cede their ecological niches to other species. The complexity of causes might explain the strange fact that some species (the dinosaurs) died out, while other species that one might think of as similar (such as crocodiles) survived and flourished.

*　*　*

Meteor impacts are not the only cosmic events that alter the evolutionary flow of life on a planet. Gamma-rays are high-energy radiation, and can be released in bursts by energetic celestial events. They interact with any ordinary material: a thin layer of material can absorb a weak beam of gamma-rays, but many gamma-rays in a stronger beam may survive during their passage through a thin layer. Gamma-rays can be used as a sterilizer in food processing, killing any bacteria that may be present. If a burst of high-enough intensity happens in the vicinity of a planet, it can sterilize it – or at least its surface, since gamma-rays cannot penetrate thick, solid material. The cave dwellers may survive such radiation, as too may the creatures living on the far side of the planet (the burst is brief and passes quickly, before the planet can complete a rotation). A burst that is less energetic may irradiate what life there is and alter its genetic make-up without actually killing things. This will dramatically alter the course of evolution of the life on the planet.

Gamma-rays are released by a supernova. A supernova is an event in which a star rapidly collapses to form a neutron star or a black hole, or in which a white dwarf collapses to become a neutron star, or two neutron stars merge to make a black hole. Supernovae occur every thirty years on average in our Galaxy. Most happen at such distances that the burst of gamma-rays is much diluted and is harmlessly absorbed at the top of the Earth's atmosphere. But it has been estimated that the Earth has been struck by strong-enough bursts of gamma-rays to irradiate its surface many times in its 4,600-million-year history. It is not known whether any of these bursts have been responsible for any specific mass extinctions, but they will certainly have had an effect on the evolution of life. According to astronomers Arnon Dar, of the Technion Space Research Institute, Israel, and Ray Norris (b. 1953), of the Australia Telescope Facility in 2002, the effect of the initial gamma-ray burst from a nearby supernova would start on the side of Earth facing the explosion, and shock waves would rip into the atmosphere. The temperature of the atmosphere would then rise rapidly, causing storms and hurricanes. All organic material on the facing surface of Earth would burn. After the gamma-ray burst would come cosmic rays, which irradiate the planet for days with lethal doses penetrating even hundreds of metres into the ground and into the sea. Statistics suggest that this will happen to a given planet once every 100 million years or so. This is about the rate of global extinctions on Earth.

* * *

We must not think of all these cosmic events as 'disasters'. I suppose that we are naturally inclined to consider an alteration of the status quo as bad news, and certainly the more popular newspapers sell in larger numbers by reporting and forecasting catastrophes, rather than by telling us how everything is going to be all right in the long run. What we can say is that the cosmic environment for life is punctuated by disasters that may cause extinctions, but that also stimulate progressive change, and consequent evolution.

Since cosmic catastrophes are sporadic events in the general flow of evolutionary development, they have helped change the way that geologists and evolutionary biologists view the growth of the Earth and life on it. At the end of the eighteenth century, decades before Darwin developed his theory of evolution, the French palaeontologist Baron Georges Cuvier (1769–1832) put forward a belief consistent with Creationist biblical interpretations, that catastrophes (such as Noah's Flood) were the causes of Earth's history, including the extinction of species. In his book *Essay on the Theory of the Earth* (1818), Cuvier wrote:

These repeated [advances] and retreats of the sea have neither been slow nor gradual; most of the catastrophes which have occasioned them have been sudden; and this is easily proved, especially in regard to the last of them, the traces of which are most conspicuous.... Life, therefore, has often been disturbed on this Earth by terrible events – calamities which, at their commencement, have perhaps moved and overturned to a great depth the entire outer crust of the globe.... Numberless living things have been the victims of these catastrophes.... Their races have become extinct.

The Russian-Jewish psychiatrist Immanuel Velikovsky (1895–1979) revived similar views in the 1950s, knitting mythology, archaeology and pseudo-science into fantastic but very popular theories about floods, cosmic fires, etc. Scientists feel instinctively uncomfortable with this worldview, called catastrophism, because everything is subject to arbitrary events – a 'tale told by an idiot' – and therefore not amenable to intellectual analysis. Reacting against such views, geologists James Hutton (1726–1797) and Charles Lyell (1797–1875) developed uniformitarianism and gradualism, the idea that geological change in fact occurs slowly over long periods of time, more along the lines of Darwinian evolution. Charles Lyell expressed his opposition to catastrophism in the following words, from his textbook *Principles of Geology* (1833):

In our attempt to unravel these difficult questions, we shall adopt a different course, restricting ourselves to the known or possible operations of existing causes; feeling assured that we have not yet exhausted the resources which the study of the present course of nature may provide, and therefore, that we are not authorized in the infancy of our science, to recur to extraordinary agents.

The Scottish geologist Archibald Geikie (1835–1924) put it rather more pithily: 'The present is the key to the past,' meaning that geologists can reliably infer what happened in the past by examining what is happening in the present and extrapolating backwards in time.

As a result of the discovery of the significant part that meteor impacts play in geology, and recognizing the chaotic nature of the gravitational effects in the Solar System that provoke them, geologists and planetary scientists now combine the two opposing views and believe that geological and evolutionary history is a mixture of slow and gradual change with sudden random jumps and occasional catastrophic events. If life finds its way not only through steady cyclical growth but also in unexpected fits and starts, perhaps it found a means, on Mars, to survive the disaster that struck the planet, as we shall see in the next chapter.

Captions to the images begin on p. 169.

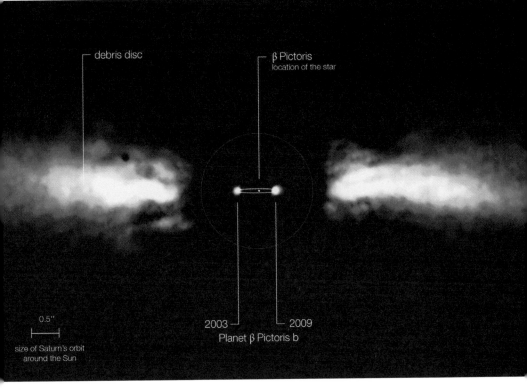

debris disc

β Pictoris
location of the star

2003 — — 2009
Planet β Pictoris b

0.5''

size of Saturn's orbit
around the Sun

1

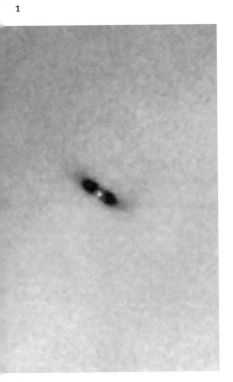

2

3

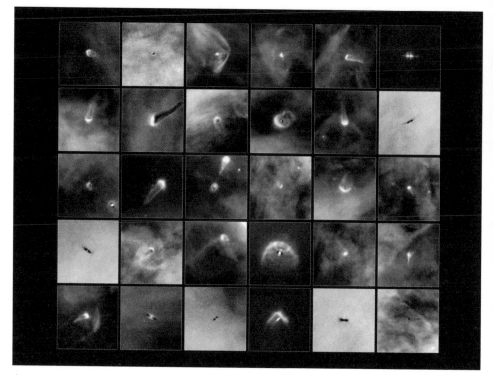

4

5

Wow!

```
    1           2              1     4     ?
    1    16     1        1           1
    1    11     1        1           11    1
         1                     3           1
         6  2                  31
    1  E 24    3    12    1     21    1
         Q  1    6  1    2    1    1         1
         U  31        1        3    7    1
    2  J  1    31   3   111        11   1    1
         5  1                1         1
         14        1        113        2     11
    1    3    1        1        1
    1    4             1        1    1    11
         4        1    1    1    11        111
         1                  1        2    1
    1    1    1             1         11    1
    1              1                  14
```

6

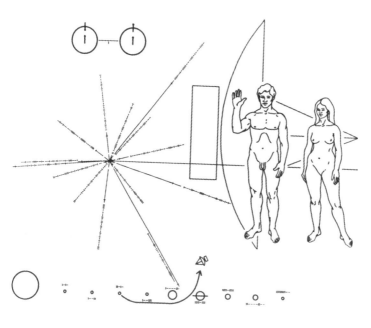

7

8

9

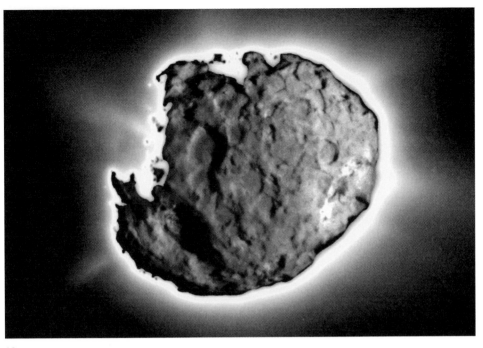

10

11

157

12

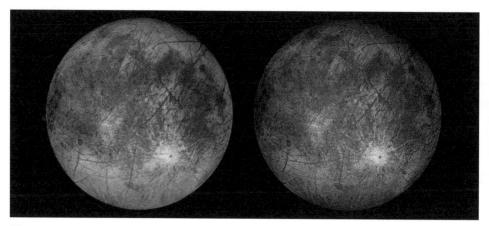

13

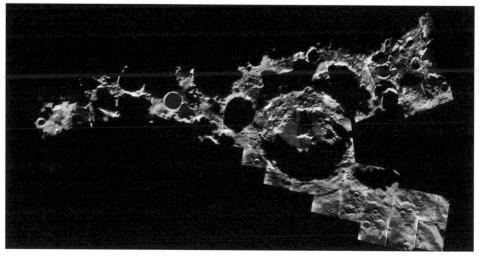

14

15

16

17

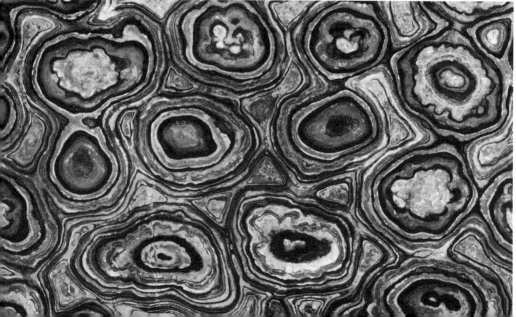

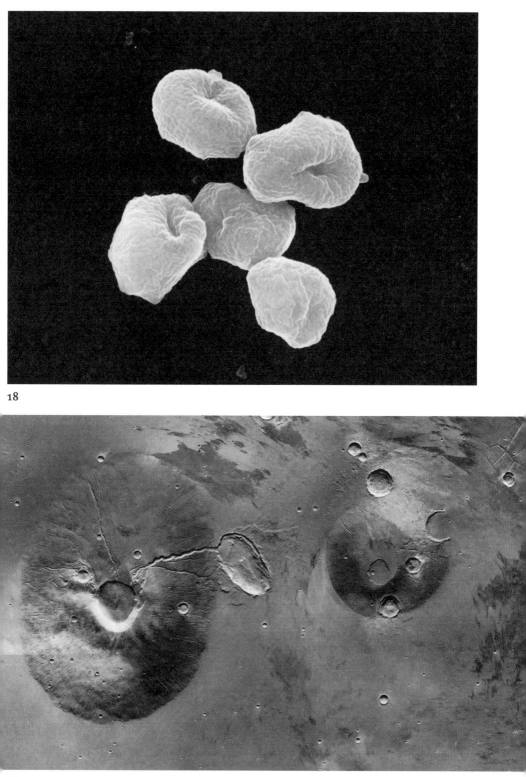

18

19

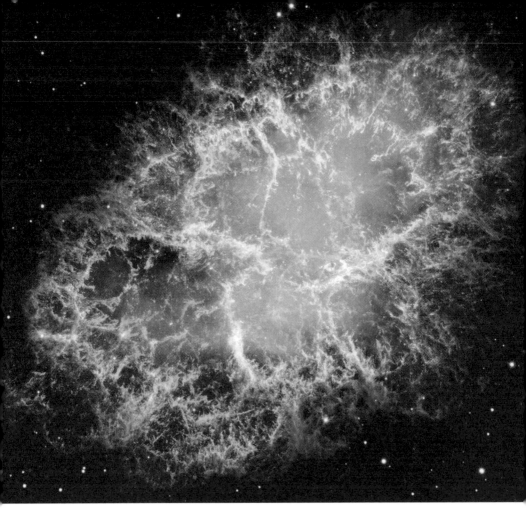

20

21

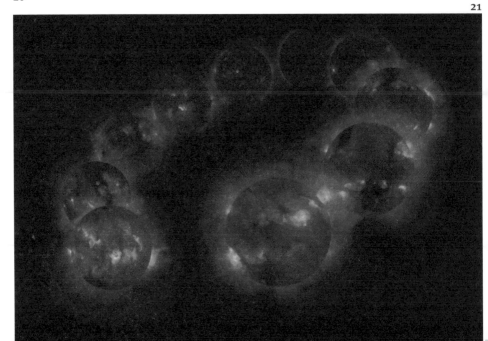

22

23 24

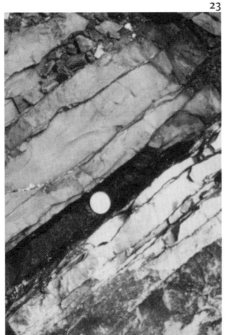

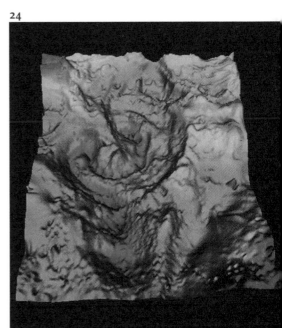

27

28 29

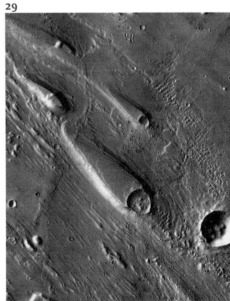

2/3"

2/3"

30

31

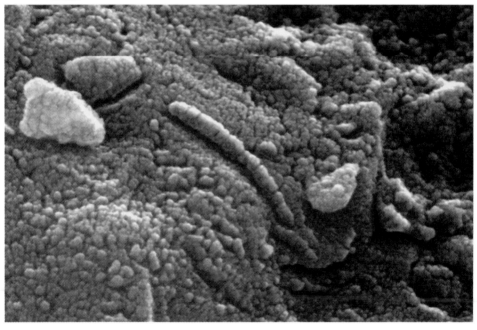

32

33

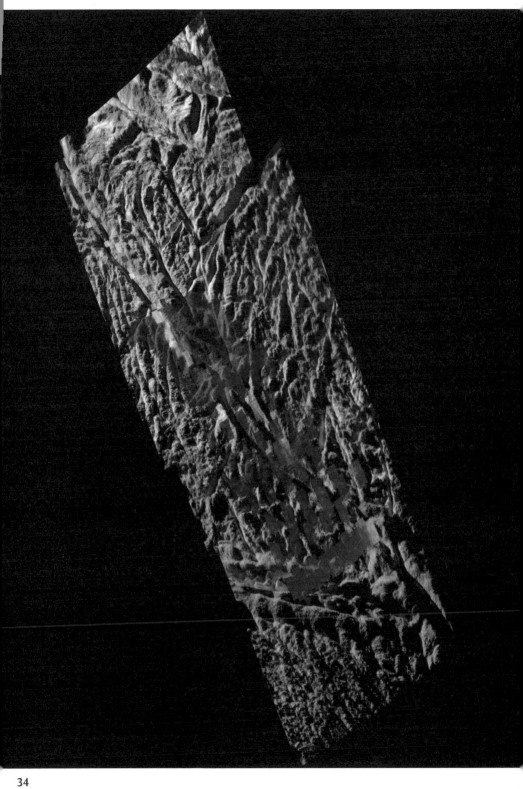

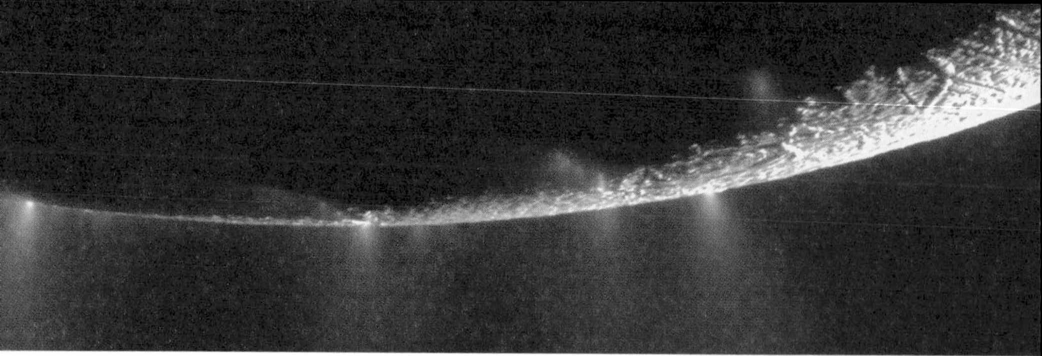

35

36

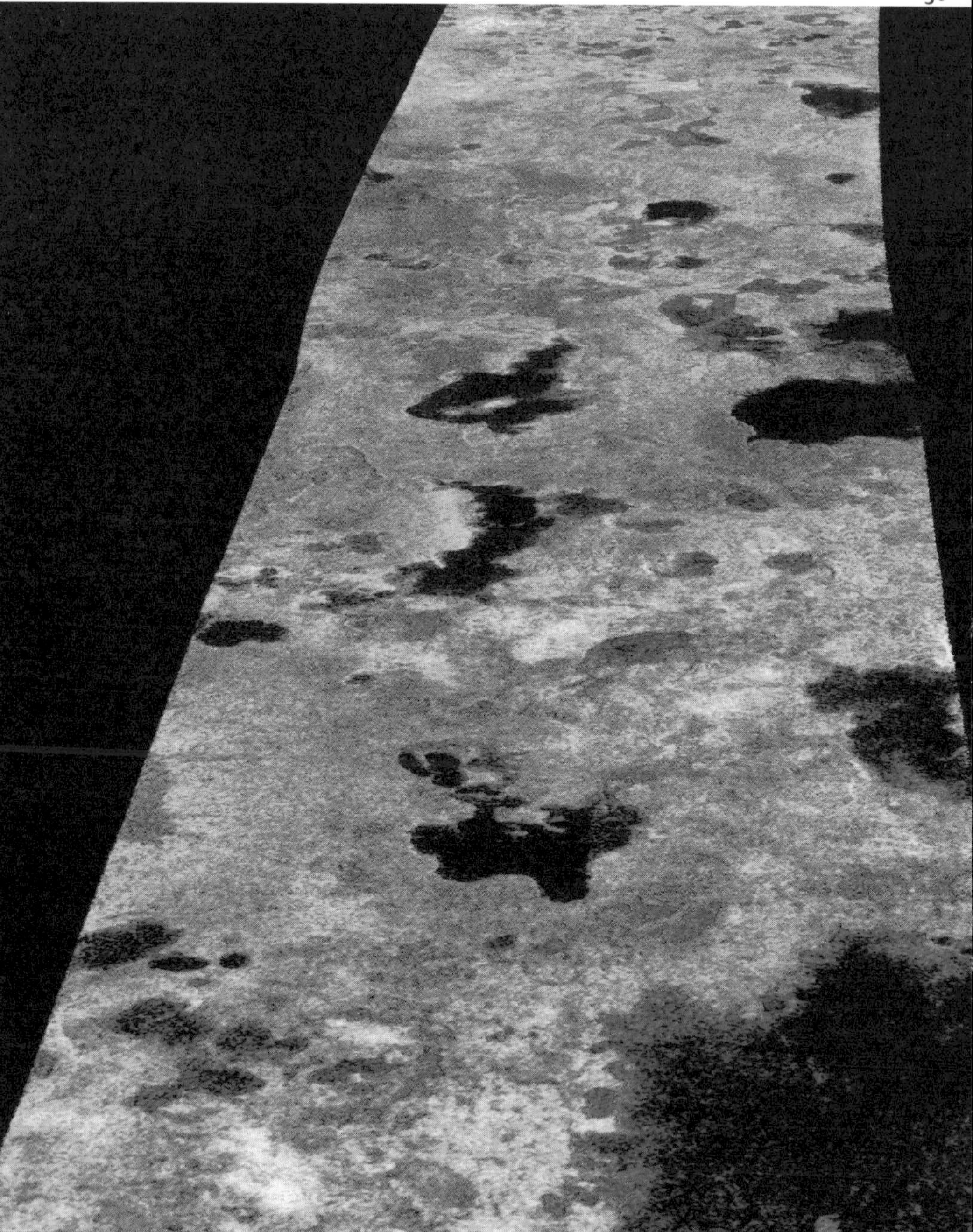

Captions

1 The planet orbiting Beta Pictoris

The blurry image of the star Beta Pictoris has been hidden at the centre of this picture of its surroundings, and replaced by a drawn dot showing its location. This double-exposure image of Beta Pictoris was taken by the Very Large Telescope of the European Southern Observatory at Paranal in Chile. The first exposure was taken in 2008, showing a planet, Beta Pictoris b, on the left. The second exposure, from the autumn of 2009, shows the planet after it had accomplished half its orbit and moved to the right. The planet orbits in the central hole of a disc of dust that surrounds the star edge-on, illuminated by starlight as a wispy cloud of material; the dust is what remains of the material from which the star and its planet formed.

2 A protoplanetary disc in Orion

The Great Nebula in the constellation Orion forms an illuminated backdrop against which is silhouetted a 'proplyd', the protoplanetary disc of dust that orbits a newly formed star. The star is buried in the dust and its light is absorbed, but this picture obtained with the infrared channel of the Hubble Space Telescope shows the star's radiation shining through an inner hole in the tilted disc. Most, if not all, stars are born this way, with the potential for a planetary system to form from the dust.

3 A star nursery

Young stars and planetary systems are forming in the nebula NGC 6729. The baby stars are veiled in this picture, behind the dust clouds at the upper left of the picture, but other stars in the area have illuminated the dust and gas in the vicinity and caused these beautiful shapes and colours.

4 The birth pangs of planetary systems

The protoplanetary discs of the Orion Nebula show a variety of forms in this mosaic of examples. Some are symmetrical, but others are distorted, buffeted by dynamic flows of gas in the heated nebula – these planetary systems will probably be still-born.

5 The Milky Way

The stars of the Milky Way arch above the European Southern Observatory in Chile. The massed suns of the Milky Way number in their billions, and the number of planets hidden in their glare is equally large. Looking at such a picture, most people feel it is inconceivable that the Galaxy is sterile.

6 Wow!

This computer printout shows, line by line, top to bottom, ten-second period by ten-second period, the intensity of radio transmissions from a random point in the sky that was passing over the Big Ear radio telescope. Each column represents a frequency channel in the spectrum of the sky near the 1,420 MHz frequency. Because of limitations in the display capabilities of the computer equipment of the 1970s, and the better to show the fainter intensities (in the expectation that any signal would be faint), the intensity is coded by a single-digit character. The intensity is a number, 1 to 9, or if the intensity is more than 9 units, it is represented by a letter up to Z. Intensities of one, two or three times the background intensity are common, due to random fluctuations of the signal, but in just one spectral channel (second from the left), for a period of a minute, the intensity was abnormally high: 'U' represents an intensity of thirty times the background intensity. Radio astronomer Jerry R. Ehman highlighted this narrow-band burst with a red pen and scrawled a surprised comment in the margin of the printer paper, 'Wow!' This is the best evidence yet, although inconclusive, for a radio signal from extraterrestrial intelligence.

7 The Pioneer plaque
The Pioneer 10 spacecraft is the
first manmade object to travel
from the Solar System into interstellar
space. It now lies at the edge of the
Solar System, more than a hundred
times further away from the Sun than
the Earth is. It carries this pictorial
plaque, designed to inform any
scientifically educated extraterrestrial
beings who find it about when and
where the spacecraft was launched,
and by whom.

8 The Arecibo message
The Arecibo message was sent to the
globular cluster of stars M13 in the
constellation Hercules in 1974. It is
a digital picture of 73 lines of 23 pixels.
Colour is used in this representation to
highlight the icons embedded in the
picture, which are intended to identify
the senders of the message to anyone
who receives it.

9 Our unique world
A pilgrim has travelled to the edge of the
Earth and looks through the celestial
spheres to see how they rotate. Within
the sphere of stars exists only one world,
ours; human beings must be unique in
the Cosmos. The illustration purports to
be a German medieval woodcut, but is in
fact a fake, concocted by the French
popularizer of astronomy Camille
Flammarion in the early twentieth
century to illustrate the Ptolemaic
view of the Universe.

10 Vapour jets from Comet Wild 2
Comet Wild 2 is about 5 km (3 miles) in
diameter. To create this composite image,
a short exposure from 2 January 2004,
which exhibits the comet's surface
craters, was overlain on a long exposure
that shows fountains of ice, dust and
vapour jetting from vents in the surface.
Comets act as a delivery system for water
and organic molecules, distributing them
·and their potential for life throughout
the Solar System.

11 Comet McNaught
In 2007 Comet McNaught was probably
on its first visit into the inner Solar
System, leaving material from its
magnificent tail to fall on planets,
perhaps seeding them with the
biochemicals that may evolve into life.

12 The asteroid Vesta as a meteor source
The south pole of the asteroid Vesta,
imaged in 2011 by the Dawn spacecraft,
shows a gigantic crater – probably caused
by collision with another asteroid. Some
of the material scooped out from this
hole by the collision falls every year as
meteors, onto planets including Earth,
moving material from one place to
another. This could be a way that life
has been propagated throughout the
Solar System.

13 The icy surface of Europa
Jupiter's ice-covered satellite, Europa,
imaged by the Galileo spacecraft in
1996. The left image shows Europa's
approximate natural colour. The image
on the right enhances colour contrasts.
Dark-brown areas represent rocky
material on the surface, and blue plains
at the two poles are ice: coarse-grained
(dark blue) and fine-grained (light blue).
Long, dark lines are fractures in the ice
crust, stained by briny material welling
up from below, possibly including the
remains of creatures from the ocean
below the ice surface. The bright feature
containing a central dark spot in the
lower third of the image is a young
meteor crater called Pwyll (after the
Celtic god of the underworld).

14 The icy south pole of the Moon
At the south pole of the Moon lie
craters with high walls, the bottoms of
which never see sunlight. In these areas
of perpetual night, frozen ice binds the
dusty soil – a possible source of water
for human life, were we ever to colonize
the Moon. This mosaic image was
obtained by ESA's SMART-1 spacecraft
in 2006.

15 Strata of the Grand Canyon
The horizontal upper strata of the Grand
Canyon contain fossil remains from the
recent history of complex animals on
Earth, and record the climate in which
they lived. The lower strata, just above
the surface of the Colorado River visible
at the bottom of the canyon and the
picture, raked at an acute angle by the
folding motions of tectonic plates,
date back to the epoch of unicellular
organisms more typical of life on this
planet than the fossil remains that lie
at the top of the cliffs. The canyon wall
represents a third of the history of
the Earth.

16 The Moon's Genesis Rock
Astronaut David Scott noticed the
Genesis Rock because of its colour –
lighter than those around it – while
exploring the Apollo 15 landing site on
the Moon in 1971. He photographed the
rock in situ, with a colour-calibration
scale in the field of view. Later analysis
proved it to be almost as old as the Earth.

17 Fossil stromatolites
Cross-section through a rock in the
Barbeton Greenstone Belt, South Africa.
The circular structures are ancient
fossilized stromatolites (see pp. 76–78).
The rock is quartz, up to 3,500 million
years old, with inclusions of kerogen
(black layers), a fossil-fuel-like deposit
of organic origin. Each of the circular
layers between began as a covering
of blue-green algae (cyanobacteria),
which captured grains of sediment
and became mineralized, building
up to a stromatolite, or rocky mound.

18 Archaea
A high-magnification image of a group
of archaea, *Sulfolobus archaea*. Sulfolobus
is an extremophile that is found in hot
springs and thrives in acidic, sulphur-
rich, high-temperature environments.
These are modern descendants of the
first forms of life on Earth, created in
'black smokers' (see p. 93) on the seabed.

In real life, these archaea are 20,000 times
smaller than this reproduction.

19 The Rahe crater on Mars
Ceraunius Tholus and Uranius Tholus
are two volcanoes located in the Tharsis
region of Mars, topped by volcanic
craters, rising up to 5,500 m above the
surrounding plain. Icy clouds drift past
the summit of Ceraunius (on the left). Its
flanks display deeply incised valleys, of
which the longest and deepest is about
3.5 km wide and 300 m deep. The valley
terminates with a fan of deposited
material in the north of the elliptical
crater Rahe. The crater itself was created
by the oblique impact of a meteorite,
which in turn may have been the origin
of the SNC meteorites, which fell to Earth
from Mars.

20 The Crab Nebula supernova remnant
The Crab Nebula is the remains of a
supernova explosion that occurred in
AD 1054. It is a shell of out-flowing gas.
The shell contains the remains of the
star that exploded and both the
elements that were made inside the
star before it exploded and elements
made during the explosion. These
include carbon and the other elements
necessary for life.

21 The sunspot cycle
Images from the Japanese Yohkoh
satellite, which is sensitive to X-rays,
depict the Sun year by year and show
how solar activity varies during a sunspot
cycle. At its most active (centre, front)
the Sun emits deadly clouds of solar
cosmic radiation.

22 The active Sun
The Sun both sustains and threatens life
on Earth. It supplies the energy used by
life to develop, and particles of solar
radiation that would irradiate and kill
living organisms if it were not for the
defensive properties of the Earth's
magnetic field.

23 The K-T boundary
At this site, near Gubbio, Italy, the K-T boundary is a clay layer (shown here marked by a coin) that separates two strata of limestone. The clay contains iridium, and is composed of material that fragmented and was distributed worldwide after an asteroid impacted in Chicxulub, Mexico, 65 million years ago – an event that contributed to the extinction of the dinosaurs.

24 Chicxulub crater
A computer-generated gravity map image of the Chicxulub crater, located on Mexico's Yucatan Peninsula, shows a buried meteor crater from the impact 65 million years ago. The crater is a multi-ring basin, with the newly revealed fourth outer ring, about 300 km in diameter, making the crater even larger than had first been suspected. This is the impact implicated in the climate change from the Cretaceous period to the Tertiary.

25 The planet Mars
Frosty white water-ice clouds and swirling orange dust-storms drift over the desert landscape of Mars in this image, taken by the Hubble Space Telescope in 2001. Grey-green markings stain the desert, and it is easy to see how they could have been interpreted as zones of vegetation. One large dust-storm system churns high above the northern polar cap (seen at the top of the image), and a smaller cloud can be seen nearby. Another large storm spills out of the giant Hellas impact basin in the southern hemisphere, at the lower right. The spots left of centre are the volcanoes of the Tharsis region. In the plains in the centre of the image are the flood-carved channels that reveal the watery nature of the planet's past, the period when life might have started on Mars.

26 Mars's polar cap
The northern polar cap of Mars ends abruptly in a scarp of melting ice,

showing alternating layers of frozen ice and wind-blown dust. This is one of the reservoirs for the water that is left on Mars, which may still maintain life on the Red Planet. This image was constructed from data from the Mars Odyssey spacecraft and is an attempt to construct a realistic landscape, as a space traveller would see it.

27 'Blueberries' in the Martian desert
The Opportunity rover looks back on its route across the dry Martian desert as it inches towards the crater Endeavour. In an image in which the colours have been exaggerated, the tracks of its wheels show where they have crunched haematite spherules called 'blueberries' that litter the surface. Blueberries were made in the watery environment common on Mars at one time, long ago, before the planet dried out.

28 Water flows in gullies on the slopes of a Martian crater
At the lip of the crater Newton on Mars, streams of water flow down gullies into the crater in spring and summer. The gullies, as imaged by NASA's Mars Reconnaissance Orbiter, are a few metres across, and the salty water that flows down them stains the inside wall of the crater a dark colour. This is some of the most positive evidence that water still exists on Mars.

29 Streamlined islands on Mars
Floodwater flowed catastrophically from south to north (bottom to top) across the plain near the mouth of Ares Vallis on Mars. As shown in this image from the Mars Odyssey spacecraft, the water encountered obstacles along its path, such as the walls of craters, but scoured around them to form mesa-like islands, standing 400 to 600 m above the floodplain. This geological evidence suggests that in the past Mars had abundant water.

30 Ice on Mars

This image records how ice on Mars disappears over four Martian days (sols) in 2008. Ice was revealed at the bottom of a small trench scooped out in the Martian soil by NASA's Phoenix Mars Lander, which was the space vehicle that brought the small Phoenix Rover to the planet. In the lower left corner of the image on the left, a group of ice lumps is visible (insets). In the right image, the lumps of ice have disappeared, having 'sublimated' – a process similar to evaporation. Planetologists have long suspected that Mars was covered with subsurface ice, and this evidence proves this theory to be true.

31 Yuty, a 'splosh' crater

Lobate deposits have formed around Mars's impact crater Yuty. The deposits extend out about 40 km from the crater, which is itself 18 km in diameter. These deposits are believed to form when a meteorite impact rapidly melts ice that lies under the surface, allowing the material to flow outwards before re-freezing. This image was taken in 1977 by the Viking 1 Orbiter.

32 Fossil life on Mars?

A high-resolution scanning electron microscope image of the Martian meteorite ALH84001 shows an unusual tube-like structural form, less than one-hundredth the width of a human hair, interpreted by some scientists as a nano-fossil of a bacterium-like creature. It is located in a carbonate deposit in the meteorite, which might also be of biological origin.

33 Ice rafts on Europa

A close-up view by the Galileo spacecraft of the thin, disrupted ice crust near the crater Pwyll on Jupiter's moon Europa.

The white and blue colours outline areas that have been blanketed by a fine dust of ice particles ejected at the time of formation of the crater Pwyll some 1000 km to the south. The unblanketed surface has a reddish-brown colour that has been painted by mineral contaminants carried and spread by water vapour released from below the crust when it was disrupted. The ice surface has been fractured into icebergs, jumbled by the pressure wave of the impact.

34 'Tiger stripes' on Enceladus

A mosaic image of Saturn's satellite Enceladus, taken from NASA's Cassini spacecraft, shows pockets of heat 1 km wide in one of the fractures in the south polar region known as Baghdad Sulcus. This fracture is one of the 'tiger stripes' that erupt with jets of water vapour and ice particles from a reservoir under the surface, and is one of the most promising environments for life in the Solar System.

35 Geysers on Enceladus

From locations along the 'tiger stripes', near the south pole of Saturn's moon Enceladus, dramatic backlit plumes spray water ice – and with it, perhaps, traces of any life that has developed in the reservoir below the surface.

36 Liquid methane lakes on Titan

Radar-imaging data by the Cassini spacecraft shows large flat areas, interpreted as lakes of liquid methane, on the surface of Saturn's satellite, Titan. The lakes, darker than the surrounding terrain, are tinted here in blue. These are the locations where any precursors of the life that may develop on the moon would be found.

13

Mars: Dead or Alive?

Mars is the fourth planet from the Sun. It is the most Earth-like of the other planets of the Solar System, but it is smaller, colder and drier, and has a much thinner atmosphere. It suffered a global climatic catastrophe early in its development, and if life did indeed once develop there, its evolution must have taken a knock at a greater scale than any mass extinction that has occurred on Earth. But, at an ESA conference in the Netherlands in February 2005, which discussed the first results from the Mars Express space mission, 75 per cent of the scientists attending were reported to have thought that Mars once had life on it, and 25 per cent believed it still did. Some scientists even claim to have discovered the fossil remains of life in Martian material.

Life on Mars probably does not resemble the ubiquitous science-fiction portrait that derives from H. G. Wells's novel *The War of the Worlds*, first published in 1898. Wells based his story on the state of the planet as it was then understood. The book's narrator is an astronomer, who sees explosions on Mars that are the launches of a fleet of spacecraft. He then witnesses the Martian invaders wreak destruction on London and the south of England. The book's opening paragraph is chilling:

No one would have believed in the last years of the nineteenth century that this world was being watched keenly and closely by intelligences greater than man's and yet as mortal as his own; that as men busied themselves about their various concerns they were scrutinised and studied, perhaps almost as narrowly as a man with a microscope might scrutinise the transient creatures that swarm and multiply in a drop of water. With infinite complacency men went to and fro over this globe about their little affairs, serene in their assurance of their empire over matter. It is possible that the infusoria under the microscope do the same. No one gave a thought to the older worlds of space as sources of human danger, or thought of them only to dismiss the idea of life upon them as impossible or improbable. It is curious to recall some of the mental habits of those departed days. At most terrestrial men fancied there might be other men upon Mars, perhaps inferior to themselves and ready to welcome a missionary enterprise. Yet across the gulf of space, minds that are to our minds as ours are to those of the beasts that perish, intellects vast and cool and unsympathetic, regarded this earth with envious eyes, and slowly and surely drew their plans against us.

The War of the Worlds has been filmed in Hollywood at least three times, and in 1978 was made into one of the first 'concept' recordings by composer Jeff Wayne, with actor Richard Burton and a number of well-known and respected popular musicians. The actor and film director Orson Welles broadcast a radio drama based on *The War of the Worlds* in 1938, in his series 'Mercury Theatre on the Air'. He cast the story in the form of news bulletins, which were intended to suggest – and apparently did indeed suggest to panicked listeners – that an invasion by Martians was under way in real time. In 1996, director Tim Burton remade a spoof version of the film: *Mars Attacks!*

In all these adaptations, Martians are portrayed as hostile, warlike colonists. This concept has its roots in the historical perception of Mars in the West, as well as in any scientific understanding of the planet, current at the turn of the nineteenth and twentieth centuries, as an old planet, its environment having deteriorated to the point that it can no longer support life. Additionally, to the naked eye, Mars is clearly a red planet, a colour with psychological overtones of blood and fire. This must have been a significant reason why the planet was named after the God of War – its sobriquet even now is 'the Red Planet'. The astrological symbol for Mars is a circular shield with a spear held behind, standing up at an attacking, phallic angle – these arms, arranged in this way, have, in biology, come to signify 'male'. In medicine as practised in medieval times, governed by astrological principles, Mars presides over the muscles and the genitals. Mars is associated with strength and aggression; this characteristic was portrayed in Gustav Holst's suite *The Planets*, composed during the First World War (1914–18). In the movement presenting the planet as 'The Bringer of War', the music conveys the new horrors of mechanized tank warfare.

Astrological views such as these created a mindset that coloured the historical development in Europe and America of the perception of Mars. In the roots of the human unconscious lies an expectation about the nature of the planet Mars and its habitability, summed up in the very word 'martial', the etymology of which refers to the astrological complex of the planet's supposedly aggressive, 'macho' properties.

* * *

Over at least 200 years the expectations about Mars were grafted onto the developing scientific picture of the planet. It was first perceived as a disc, a world like ours, by Galileo. His telescope was, however, not powerful enough to see its surface markings, the first of which was discovered in 1659 by the Dutch astronomer Christiaan Huygens (1629–1695). That year he detected a shadowy, grey-green, triangular feature, now known as Syrtis Major. The Italian-born French astronomer Giovanni Domenico Cassini (1629–1712) first identified the white polar caps of Mars in 1666. In 1704, the French-Italian astronomer Jacques Philippe Maraldi (1665–1729) showed that the size of the polar caps varied throughout the Martian seasons, as if they were ice caps similar to Earth's. By 1784, in Britain, William Herschel was articulating the growing realization that Mars had a 'considerable but moderate atmosphere, so that its inhabitants probably enjoy a situation in many respects similar to ours'.

In 1840, the German banker and amateur astronomer Wilhelm Beer (1797–1850), and his colleague Johann H. von Mädler (1794–1874), made the first Martian maps, showing dark areas that seemed variable in colour and intensity. Initially these were thought to be seas, but the French astronomer and botanist Emmanuel Liais (1826–1900) suggested in 1860 that they could be large patches of vegetation, showing seasonal variations. Some astronomers thought they detected a wave of darkening of these patches, which travelled from the pole to the equator as the spring progressed. They interpreted this as the melting of the polar cap, with the water flowing equator-wards and irrigating the vegetation.

But the key observation that pointed towards intelligent life on Mars came in 1869, when it was closer to Earth than usual and could be seen more clearly. The Vatican Observatory astronomer Angelo Secchi (1818–1878) drew detailed representations of Mars at this time, which included two dark, linear features on the surface that he referred to as *canali*, the Italian for 'channels', but which is more literally translated into in English as 'canals', suggesting artificial construction. In 1877, the Italian astronomer Giovanni Schiaparelli (1835–1910) produced the first detailed map of Mars, labelling features as 'continents', 'islands' and 'bays', as well as numerous straight, linear features that he also designated as *canali*, and to which he gave the names of famous rivers on Earth. The French astronomer and popularizer of astronomy Camille Flammarion

(1842–1925) wrote that these channels resembled manmade canals, which could be used to redistribute water across the planet. The US businessman and amateur astronomer Percival Lowell (1855–1916) founded the Lowell Observatory in 1894 to investigate Mars. He published highly successful books about life on the planet, including maps with detailed networks of artificial canals that he interpreted as an irrigation system, used to carry water from the polar ice caps to the rest of the dry planet. All of this encouraged the idea that Mars might be dying, but was inhabited none the less by advanced beings.

Canali proved to be an optical illusion. The Turkish-born French astronomer Eugenios Antoniadi (1870–1944) could not see canals in the excellent viewing conditions offered by the close approach of Mars to the Earth in 1909, as seen in the very precisely figured 83 cm telescope at Meudon in France, during times of good seeing. He could see faint, blotchy structures, and, as an explanation of how the illusion came about, he cited psychological experiments that had been conducted by astronomer E. W. Maunder (1851–1928) in 1894 and 1903 on groups of schoolboys that showed how 'dark points...too small to be appreciated individually, produce on the retina the idea of diffuse lines.' Martian canals were not real. But the concept of Mars as a desiccated world proved to be close to the truth; sandstorms were observed on the planet, covering the features that had previously been identified through telescopes. The atmosphere of Mars is much thinner than Earth's, and made of carbon dioxide, nitrogen and argon – but it is relatively easy to see, with a telescope, that there are clouds, sandstorms and an exchange of material between the polar caps on a seasonal basis. During the northern winter and southern summer (when Mars is closest to the Sun) great dust-storms sometimes cover virtually the whole planet. The best views of Mars as seen from the distance of the Earth have been obtained by the Hubble Space Telescope, and show it unmistakably as a planet like ours (**25**).

Mars became much better understood through the space exploration programme. This is an international effort by such organizations as NASA and the Russian, European and Japanese space agencies. Their scientists are actively coordinating the planetary exploration programme for Mars through the International Mars Exploration Working Group, so that each space mission, no matter from which space agency, builds on the knowledge gathered in the missions before. A programme to explore a planet starts with a flyby – a quick look as a spacecraft zooms past – and then positions the spacecraft to make sustained investigations remotely from orbit. The orbiters will progressively gain capability, with better and better cameras, for example, able to view the land close up and with the potential to see changes in the planet's environment, such as may be produced by geological[1] developments or by those organisms that live there. Spacecraft will carry remote sensing equipment that will analyse the composition of the atmosphere and surface, looking for chemical traces produced by life. Probes are landed through the atmosphere and onto the planet's surface. The first might simply plunge through the atmosphere before crashing, but they will nevertheless be in physical contact with the gases of the atmosphere and able to sample them for analysis, although this will have to occur quickly! Later probes will slow their freefall with up-thrusting rockets (as in the Moon landings) and parachutes (these only work if the planet has an atmosphere, as Mars does) and land gently onto the surface. Landers can make a more prolonged study of the atmosphere by touching down softly to photograph the surroundings at close range and analyse the surface by physical contact.

Landers are stationary, and sample only one site. The high winds and global sandstorms that blow across Mars, however, mean that the soil at one site is about the same as at any other, so any one sample taken by a lander will be more or less representative of the surface of the entire planet. The next scientific capability comes from rovers, which can explore the area around the landing site. They are free to move – until they get stuck under a rock or bog themselves down in sand, or until they run out of motive power, when the solar panels that generate the power that drives them get dusty, or some other important element fails. Back on Earth, the controllers of the rover can choose the most interesting places to visit, selecting targets that look as if they will reward close inspection – different coloured rocky deposits or oddly shaped rocks, for example. The rover may be able to dig underneath the surface sand to examine niche environments specific to the site.

The exploration of Mars has reached this stage as I write, with rovers active on its surface, ranging over distances up to 35 km from their landing site, peering into a variety of Martian environments, and analysing a variety of rocks to investigate how they formed and how that process bears on the possibility that life once was or still is present there. In the future, the exploration programme will continue by returning samples to Earth for close study. These space missions are in the stage at which they are being actively planned. The scientists will select a site on the basis of remote observation; there will be much argument about where has the most promising properties, and the engineers will rule out some very interesting places as too difficult to land on! Once the site is selected, the controllers will direct a spacecraft to land there and pick up samples, packing them into a container that will be re-launched back into space, probably to be collected in orbit by the 'parked' craft that brought the lander to Mars. This spacecraft will return to the Earth and rendez-vous with a space station, so that the sample soil can be carefully brought back to a terrestrial laboratory. It will have been sealed carefully to avoid contamin-ation, and it will be at this stage – if the complicated procedure works – that we may find unquestionable evidence of life on Mars.

Manned exploration of Mars will be the next step; the scientific reason for sending astronauts is that people see more than robots, because their eyes are attached to a sophisticated processing device: the human brain. Humans pick up signs that robots miss. At this point it may be very difficult, if not impossible, to prove that if life is found it was not brought to Mars from Earth; space agencies can and do sterilize landers and rovers so that they do not contaminate the planet they visit, but we cannot sterilize people. In the distant future, colonization will be the climax of the programme. Of course, these are difficult expeditions to execute. It is at least a six-month journey to Mars, because of the configuration of the orbits. Once there, the next opportu-nity to return will occur a few weeks later or a year-and-a-half later, so the astronaut would either stay for too brief a visit, or endure one that could be too long. In any case, a round trip could last well over a year. Food and water have to be carried, if they are not recycled or grown or distilled in artificial farms. There are mortal dangers en route, including solar-radiation storms and asteroids. Only the very brave will volunteer to go to Mars for the first explor-ations, on missions that face even greater risks than the Apollo missions to the Moon (they may even, at the outset, be planned as one-way missions). None of this is easy, and the difficulty, expense and risk of each stage of the exploration programme is greater than the one before. As we shall see, about half the mis-sions sent to Mars so far have been successful in achieving their minimum planned objectives.

*　*　*

The first of what have been approximately twenty successful space missions to Mars was Mariner 4, which flew past in 1964. By today's standards its perfor-mance was primitive, but at the time it was momentous. It took a score of pictures, tape-recording them and then playing them slowly back to Earth, at a data-transmission rate approximating to that of a telegrapher of a hundred years ago transmitting Morse code. These images showed that there was nothing on Mars that resembled canals, and its surface was heavily cratered. Most of the large craters on any planet in our Solar System were produced early on in its history, at the time of the Late Heavy Bombardment. If they are still visible, as on the Moon, it means that the surface has remained constant for the last 4,000 million years. If they have all gone, as is the case with our own planet, it means that the surface is active, with weather. Of course, if the surface has been unmodified for 4,000 million years, it seems unlikely for there to be intelligent life of any significance – evidently no agriculture, towns, cities or roads have existed to domesticate the Martian landscape. Big craters, but no rain, water or life – that was the picture of Mars painted by Mariner 4.

The truth has proved to be a bit more complicated; the surface of Mars is still active in some ways. Mariner 9, which reached Mars in November 1971, was the first spacecraft to enter into orbit around another planet. Its first pictures were disappointing: a global dust-storm was in progress. All that could be seen were the south pole and the tops of four high volcanoes on the Tharsis shield volcano. Controllers waited two months for the atmosphere to clear. When it did so, they were rewarded by magnificent views of the eponymous Valles Marineris, a huge canyon that extends 4,000 km east–west along the equator. It is 600 km wide and 7 km deep. Mars has no tectonic plates, a thin atmosphere and no rain – so much of its crust is old, with some terrains having formed 3,800 million years ago. But the planet has had more recent volcanic activity and there are volcanoes, including Olympus Mons, the largest in the Solar System. It is 24 km high: as high in metres as Mount Everest, the tallest mountain on Earth, is in feet.

Viking 1 and Viking 2 were orbiters that also carried the first spacecraft to land successfully on Mars, in 1972. The landing sites were selected on the basis that they were on large plains, to increase the chances that the spacecraft would land safely and not fall over on a steep cliff-side, for example. The Viking landers found a bleak, rocky, desert landscape. The rocks were volcanic in origin, with holes where bubbles of gas had been. They were angular and fractured, having been thrown into position by the impact of meteors on the plains nearby. Where Viking 2 landed there were no rounded boulders to be seen; no rivers had ever tumbled these rocks over a streambed. The rocks at the Viking 1 landing site were similar but their sharp edges had been blunted – a subtle suggestion that the rocks might have been eroded by water.

Mars Pathfinder was the first successful mission (1997) to release a rover that could seek out interesting features near its landing site. Scientists' current knowledge of Mars comes from the more recent orbiters Mars Odyssey, Mars Express, Mars Global Surveyor and Mars Reconnaissance Orbiter, which have mapped the surface at high resolution, and the rovers Spirit, Opportunity and Phoenix, which have examined the planet's surface up close for many hours. Their operators have never seen anything that could be said to be alive, nor even to have been so in the past. Perhaps something will turn up during the explorations by the Curiosity rover, which landed on Mars in 2012. But it is possible that we are missing something – in control experiments, when geologists were shown pictures from a rover in a terrestrial desert, they overlooked fossil remains the size of dinosaur footprints.

* * *

The picture of Mars that emerges from this immense effort is of a planet that is now desert-like, dusty and desiccated at its surface – except at the poles, where there are deposits of ice and dry ice (frozen carbon dioxide), 2 to 3 km thick. It has seasons, like Earth does, and the deposits of ice at the polar caps respond in the same way, with the pole in one hemisphere contracting during its Martian summer when the ice melts, while the other polar cap expands as ice is deposited in its hemisphere's corresponding winter season. The ice is laid down after a summer season of dust-storms, so the polar caps consist of layered ice and soil deposits (26). There are numerous indications that Mars at some time long in the past had water in abundance. The Viking landing sites were unlucky choices, driven by the ease of positioning craft on the surface there. Only an imaginative reading of the subtly blunted edges of rocks at the Viking 1 landing site could have shown that in fact there do exist the geological remnants of lakes, glaciers, water running in channels with tumbled, rounded rocks on the streambeds, and massive floods. Mariner 9 was the first to picture vast, now-dry flood plains, but it was not until the Viking orbiters mapped the surface more completely that their nature was fully realized. Building on the Viking orbiters' evidence, and targeting for close-up study some of the areas where they discovered possible water erosion, recent spacecraft have imaged valleys, their sides stepped in layers of sedimentary rock, laid down in strata at the bottom of lakes or seas. Mars Global Surveyor saw not only dry rivers, but also crusty, dried-up lake beds. The Mars Exploration Rover 'Spirit' has found rocks rich in carbonate minerals in the Columbia Hills of the Gusev crater. The mineral probably comes from the chemical reaction of the original rock with water in which carbon dioxide from the atmosphere had dissolved. The abundance of carbonate minerals also indicates that Mars was warm at the time they were formed. Spirit's sister rover, called 'Opportunity', even landed by chance inside a crater, Eagle, with similar minerals laid down by the evaporation of salty water in a lake. The entire area around Eagle is littered with so-called 'blueberries', small spheres of a kind of the mineral haematite (not really blue, but certainly less red than other kinds of haematite) that forms in water (27). The Gale crater shows lots of evidence of water-deposited minerals like these, which is why it was chosen as Curiosity's landing site.

The surface of Mars has wide-ranging networks of dry valleys. Some of them are dry riverbeds, meandering at the bottom of large valleys, forming an extinct drainage system in which water has acted over a long time. Some dry valley systems have no small-scale streams in their pattern, just the main rivers, which begin full size at a particular place. This suggests that the dry beds were not made by channels of rain. They were carved by ground water flow, rivers flowing at first inside the ice glaciers, and then on the ground beneath them. There are a number of cold-weather and glacial features on Mars that indicate that there was an icy landscape there at one time, including thermokarst, pingos, polygonal ice wedges, eskers, drumlins (or hoddy-doddies) and glacial moraines.[2]

Other valleys appear to have been made by the collapse of the roof structure of an underground channel. Perhaps these are associated with geo-thermal springs, the heated permafrost melting and undermining the surface. Drainage from melting water has been imaged by the High Resolution Imaging Science Experiment (HiRISE), a camera on the Mars Reconnaissance Orbiter (**28**). Thousands of narrow, dark streaks appear on the surface of Mars in the spring, and grow in length by many metres per day during the summer. They appear on the steep slopes of some Martian craters that face towards the equator. They disappear as winter comes. The most likely reason is that the stains are caused by briny water seeping downhill in small gullies.

At the present time, scientists get excited about small seepages and stains of water on Mars. In the wet eon of Mars's history, there were large areas of standing water: seas or lakes. It is called the Noachian period, ending 3,700 million years ago. It is named after Noachis Terra (Noah's Land), the region of Mars that shows the most extensive geological features that reveal the action of water, itself named after the biblical survivor of the Flood. Rocks

Table 6. The Geological Timescale on Mars

Period	Ended (millions of years ago)	Began (millions of years ago)	Age of Mars (millions of of years)	
Amazonian	0	3,000?	1,500?–4,530	
Hesperian	3,000?	3,700	800–1,500?	Production of lava plains Global climate change Catastrophic floods
Noachian	3,700	4,100	400–800	Glaciers and river valleys The Late Heavy Bombardment
Pre-Noachian	4,100–	4,530	0–400	
		4,530	0	Formation of Mars

from the Noachian period survive in abundance on Mars, although those of the same age on Earth, from the Archaean eon, are rare: plate tectonics fractured continents from this period and buried most of the rocks (chapter 8).

Not only was there water on Mars that was running over geologically significant periods of time, but there were also massive floods that individually lasted for a matter of only weeks, and produced temporary lakes. There are stepped cliffs on the interior walls of craters, wave-cut platforms formed at different levels of the water that once filled them. This was the 'Paradise' of Martian history, called the Hesperian era, after Hesperia Planum, the typical terrain of intermediate age on Mars. How long it lasted is a matter of debate – at least 500 million years and perhaps a lot more.

A startling discovery by the Viking orbiters was of crater-crowned 'islands', standing proud above a dry plain at Ares Vallis in the Chryse Planitia region (**29**). Their lozenge shapes suggested that they were made by a flood of more-than-biblical proportions. The flood formed streamlined islands around craters 10 km in diameter. The height of the cliffs that surround the islands is 400–600 m. The flood that carved these cliffs must have been on a catastrophic scale. Mars Pathfinder saw varied, rounded rocks and boulders, brought from great distances by the surge of another flood. These events would have been truly awesome, releasing perhaps 100,000 cubic km of water in a matter of a day or a few days – by comparison, a typical flood on Earth may be just a few cubic km, with the largest known from the geological record amounting to 100–1,000 cubic km. Such massive floods were caused by the collapse of a natural dam. On Mars the dams were likely to have been made of ice. They were eaten away by evaporation after the Noachian era, honeycombed and weakened to the point of collapse under the pressure of the pent-up water behind.

For the purpose of scientific analysis, the geological periods on Mars have been divided up in a manner similar to those of Earth. Of course there is much less in the classification. The dates that divide the different periods are very approximate, based on counting the density of craters on plains of rock of various kinds and matching this with what we know of the density of craters in rocks of different ages on the Moon. The result is **Table 6** (p. 182). When compared with **Table 1** (p. 69), which charts the Earth's history, the classification of geological periods on Mars obviously lacks detail, but of course no geologists have yet taken their hammers to excavate the rock strata of the Red Planet.

* * *

Is there water on Mars still? When the Phoenix Lander dug into Martian soil in 2008, it exposed an evaporating white deposit (**30**). This showed dramatically the conclusions drawn from other more indirect evidence: there are quantities of ice below the surface of Mars. Some of it is at the poles, in the ice caps, but there is much ice everywhere on the planet, under the surface. Some recent meteorite craters – the one known as Yuty, for example – are surrounded by out-flowing lobes that look like the petals of a flower. These patterns are not found on the Moon or on Mercury. Such craters as Yuty are known as 'splosh' craters (31), and the lobate pattern is identified as a 'fluidized impact surface pattern', as if formed by projectile water, impacted in mud. This suggests indeed that the subsurface of Mars was ice, and that it was melted by the impact of the meteor, flowing outwards and re-solidifying.

In fact, it seems likely that much of Mars is covered in a permafrost many metres deep. The Mars Odyssey spacecraft carried a gamma-ray spectrometer in order to determine the composition of the surface by remote sensing. It discovered that there was a frozen ocean of icy water an area about 5,000 km wide in the southern hemisphere. It is at least a metre deep and maybe deeper, and contains enough water to fill one of the Great Lakes on the Canadian–US border.

The permafrost is the source of water seen by Mars Global Surveyor, sporadically staining the Martian ground as water ran down inside the cliffs of crater walls. Geothermal activity seems to have been responsible for melting some of the underground ice, which seeped along underground channels and now drains intermittently from springs.

It seems possible that the past of Mars may have developed life, and some limited areas have the conditions to sustain it even now. The dusty Martian atmosphere is mainly comprised of carbon dioxide and nitrogen. Intriguingly, however, as discovered by the Mars Express orbiter in 2004, there are also traces of methane. On Earth this gas is emitted by active volcanoes (there are none known on Mars) and by biological activity (the gas's common name is 'marsh gas', because it is emitted by rotting vegetation, but, as we have seen, it is also emitted by animals). Because the gas is so unstable, it does not last long even in Mars's thin atmosphere – so something there must be continuing to produce it. Whether the sources of the methane are unspectacular volcanic vents or flatulent bacteria remains to be proven.

* * *

What happened to Mars to make it so dry after such a wet past? Measurements made by the first Mars probes revealed that, surprisingly, Mars does not have an extensive magnetic field – although in 1999 the Mars Global Surveyor did discover residual magnetism in the southern hemisphere, in a regular pattern of stripes with alternating polarity. The magnetic stripes were laid down in old surface materials as the convective iron core of the planet formed and re-formed under the surface. At some time about 4,000 million years ago, at the end of the Noachian period, the core solidified and the Martian dynamo died away, together with the magnetic field that it generated.

The magnetic field of a planet defends its atmosphere from the action of solar particles that emanate from the Sun (chapter 10). Earth's magnetic field deflects these particles, but since Mars now has no protective magnetosphere, its atmosphere is exposed directly to the solar wind, and much of it has been stripped away. The loss of its magnetic field caused the atmospheric pressure of Mars to drop, and the water and ice on the planet's surface to evaporate – most of it was lost to space, but some seeped below the ground and froze as permafrost. Not only has Martian life been deprived of its abundant water and benign atmosphere, but also now ultraviolet light and solar particle radiation strike its surface and would be deadly to any life that dwells there. If life did exist in the past, it must have migrated to more sheltered places, such as caves and underground chasms; one calculation suggests that the radiation intensity is so high that life could survive and develop only at depths greater than 7 m below the rocky surface.

The loss of the atmosphere also caused a drop in temperature on Mars. The planet's greater distance from the Sun means that it is colder than the Earth. The weak atmosphere makes the situation worse. The atmospheric pressure is 1 per cent that of Earth's, too low for liquid water to survive even if there is any. The temperature at the equator on Mars on one of its warm summer afternoons reaches only about 20°C, and drops to freezing point at sunset. The weak greenhouse effect and lack of atmospheric blanketing mean that the heat of the Martian day escapes readily into space as the Sun goes down, and the night-time temperatures are bitterly cold. On the equator in summer the temperature falls to –70°C at night, –140°C in the polar regions. Any life on Mars that survives will have needed to adapt to these extremes of temperature variation.

* * *

If Mars is not an entirely dead world, it must be nearly so, having suffered climate change beyond anything our Earth has experienced. The Viking missions of 1976 looked to see whether any signs of life remain. Apart from analysis of material brought back from the Moon, these are the only experiments carried out on another planet to see if it contained life. The Viking landers undertook four experiments, all of them repeated at each of the two sites. They searched for life in spoonful-sized samples of soil taken from the surface, using scoops that dug into depths less than 10 cm or so. The designs of the experiments were based on assumptions about what life might be present in the soil, presupposing carbon-based life similar to that found on Earth.

Pyrolytic Release experiment (PR): This experiment on the Viking landers looked for radioactive compounds fixed in soil that had been exposed to radio-active carbon dioxide. The idea was that any Martian micro-organisms would have needed to be able to breathe in carbon dioxide from the atmosphere and convert it, without water, to their organic body-matter, as plants do on Earth, by such a process as photosynthesis. A spoonful of Martian soil was put in a chamber with carbon dioxide that had been labelled with radioactive carbon-14. It was incubated and lit for 120 hours by an artificial sun (a xenon arc lamp), to provoke a reaction akin to photosynthesis. The atmosphere was vented away. The soil was then heated to about 625°C to break down ('pyrolize') any organic matter. If the experiment had worked, and organic gases were indeed released, they would be detected by a Geiger counter. As controls, other samples of soil were first heated and sterilized before being exposed to the same tests: if these sterilized samples performed in the same way as the unsterilized ones, the changes must be some complicated inorganic chemistry, not Martian microbes. Very small amounts of carbon from the radioactive carbon-dioxide atmos-phere did become part of the soil, but this occurred in both the unsterilized and the sterilized samples, and also in those exposed to artificial sunlight as well as those that were not. All of this suggested that the effect had been the result of something non-living, and had certainly not been achieved by any process resembling photosynthesis.

For both of the next two experiments to work, the micro-organisms would have to be similar to terrestrial ones, and to feed on similar food, washed down with water.

Gas Exchange experiment (GEX): This experiment tested for changes in an atmosphere caused by micro-organisms that might be present in the Martian soil. The idea was that the soil might contain organisms that had been dried out and dormant for a very long time. Just adding water might reactivate them. A spoonful of soil was put in a small chamber and made wet under an atmos-phere that mimicked that of Mars. Its composition was repeatedly measured over ten days to see if the sample emitted carbon dioxide. A further idea was to provide the micro-organisms with a broth of organic nutrients, to encourage them to grow. What happened was unexpected: oxygen was produced when the soil was made wet, although oxygen had never before been seen in tests on terrestrial and lunar soil. When the nutrients were added there was indeed a slow trickle of carbon dioxide, as expected if there was life in the Martian soil, but the unknown process that produced the oxygen is a confusing circumstance, and makes it doubtful that biology was involved in producing the oxygen and carbon dioxide. They must have been produced by some ordinary chemical processes on the chemicals of the Martian soil, although the exact chemical reactions have not been identified.

Labelled Release experiment (LR): This experiment looked for the release of radioactive carbon dioxide by metabolism by any micro-organisms in Martian soil from organic material that had been labelled by radioactive carbon-14. A sample of Martian soil was placed inside a culture chamber and fed a broth of organic nutrients that had been manufactured with radioactive carbon-14. The expectation was that if micro-organisms were present in the soil sample, they would consume the nutrient and breathe out radioactive carbon dioxide, detected with a Geiger counter. There was indeed a rapid release of carbon dioxide, followed by a prolonged slow release.

The LR experiment on the Viking landers was the most exciting, because its outcome almost exactly matched the expectation that had been identified before the launch as indicating the existence of extraterrestrial life. The initial speed of the reaction, however, suggested that the outcome might have been caused by an oxidizing chemical, such as hydrogen peroxide, rather than something biological. The scientists who analysed the Viking experiments overall decided that they had not been strict enough before the launch in enunciating the meaning of the various possible outcomes. They were particularly influenced by the fourth experiment:

Gas Chromatograph Mass Spectrometer (GCMS): This instrument was used to search in Martian soil for organic compounds usually found in terrestrial life. Soil was warmed to make any carbon compounds turn into gases, and then the instrument simply looked to see what was there. The biggest surprise was that it found no carbon compounds at all – less carbon in organic compounds than had been found on the Moon (probably deposited there by meteorites).

The results of the Viking experiments have stood the test of time, but more than thirty years later scientists are still arguing about precisely what they mean. Things happened that could have been organic but were not always what was expected – the 'wrong' gases were given off, for example, or the results did not repeat in the ways predicted.

For a long time, the key observation was thought to be the complete absence of any organic compounds in the Martian soil. This suggested that something complicated but inorganic was going on, although precisely what that was remained unclear – the most probable cause was thought to be that hydrogen peroxide had been formed by ultraviolet light acting on the soil, and that this had given rise to all the false positives. As the Viking project scientist Gerald A. Soffen (1926–2000) put it, the chemical experiments revealed a surprisingly active surface on Mars, but 'we can't find any bodies!': no dead bacteria. For a long time this result cast a pall over the expectation that there might be life on Mars – a depressing thought for the science of astrobiology.

The lack of bodies became less startling, however, when in 2008 the Phoenix mission discovered perchlorate chemicals in the Martian soil. Apparently, like the previously hypothesized hydrogen peroxide, these had been produced by ultraviolet light. Perchlorates are inert at Martian temperatures, but when warmed become highly reactive with carbon, and the GCMS (Gas Chromatograph Mass Spectrometer) soil samples were prepared for analysis by warming. It is possible that the reason why the GCMS saw no organic material is that just before the soil's analysis any organic material had combined with the perchlorate chemicals. Effectively the soil had been bleached as clean as a sink in a kitchen-cleaner advertisement on television.

The results of the Viking experiments were confusing, because Mars did not conform to the scientists' expectations. The consensus is that none of the experiments shows that there is currently life on the surface of Mars. On the other hand, they can also be interpreted as consistent with the presence of life together with an unusual surface chemistry. It remains a common hope among scientists that life did develop in the wet and warm era of Martian history, and survives in niche environments to this day, presumably underground. The most dramatic, but also the most controversial, indication that there was life on Mars is the discovery in 1996 of alleged biomarkers, including fossil bacteria, in a Martian meteorite.

* * *

The best place to discover meteorites on Earth is Antarctica. They are relatively easy to find on the white surface of ice, with the nearest terrestrial rock 900 m (3,000 feet) below. Meteorites are concentrated on the surface because ice flows, as is shown by the existence of glaciers. Although ice is normally reckoned to be solid water, a layer of ice standing on solid rock can be melted a little at its base by the high pressure of the weight of ice above, so it can slide along. If ice is confined in a bowl-like depression in the ground, fed by a glacier at the head of the bowl, overflowing at the lowest edge, the flow churns up meteorites that have fallen over time. Meteorites that were buried at the bottom of the bowl are brought to the top of the ice layers and exposed when the surface ice evaporates. The first Antarctic meteorite was discovered in 1912, by a member of Australian geologist Douglas Mawson's (1852–1958) exploratory expedition, but in 1969, Japanese glaciologists discovered nine meteorites within 3 km of each other – they were of five different types and therefore not fragments from the same fall. This brought home the importance of Antarctica as a place to find and analyse meteorites. The Japanese National Institute of Polar Research and the University of Pittsburgh set up expeditions in the mid-1970s, which led to the establishment of the Japanese Antarctic Meteorite Research Center in Tokyo, and the US Antarctic Search for Meteorites Program (ANSMET). Tens of thousands of meteorites have since been collected.

Roberta (Robbie) Score of the ANSMET team discovered ALH84001 (its number signifies that it was found in the Alan Hills ice field in North Victoria Land in 1984). She was immediately struck by its unusual green colour. Its composition and mineral content are similar to the SNC meteorites, thought to be from the planet Mars; in fact, one of them, EET79001, has samples of gas trapped in bubbles within, the composition of which definitively matches measurements made on Mars by the Viking lander spacecraft.

ALH84001 was ejected from the surface of Mars by the impact of an asteroid. The history of the meteorite is as follows. The original igneous rock solidified within Mars about 4,500 million years ago. Between 4,000 and 3,600 million years ago, in the Noachian period of Mars's history, the rock was fractured. Water then permeated the cracks, depositing carbonate minerals. These claims are not contentious, but the next part of the story is. According to a group of scientists in 1996, led by Everett Gibson and David McKay of NASA's Johnson Space Center, bacteria colonized the rock, living in the fractures and leaving fossils. The story of the rock returned to uncontroversial episodes 16 million years ago, when a large meteorite struck Mars, dislodging the rock and ejecting it into space. After orbiting in space, the meteorite then landed in Antarctica about 13,000 years ago, where it stayed until its discovery in 1984. It was numbered ALH84001 at the time it was put into storage, but a labelling mix-up resulted in ALH84001 being wrongly classified as a diogenite from the asteroid Vesta (see p. 86). It attracted little attention until 1993, when, seeking to research diogenites, a scientist from the Lockheed Corporation, David Mittlefehldt, looked at the rock and recognized it to be a new kind of meteorite, very like the SNC meteorites. It was Martian, not Vestan.

The analysis by NASA scientists of ALH84001, and their conclusion that it contained fossilized Martian bacteria, attracted high-level attention. Speaking about the analysis, US President Bill Clinton said, in 1996, on the lawn of the White House:

Today, rock 84001 speaks to us across all those billions of years and millions of miles. It speaks of the possibility of life. If this discovery is confirmed, it will surely be one of the most stunning insights into our Universe that science has ever uncovered. Its implications are as far-reaching and awe-inspiring as can be imagined. Even as it promises answers to some of our oldest questions, it poses still others even more fundamental. We will continue to listen closely to what it has to say as we continue the search for answers and for knowledge that is as old as humanity itself but essential to our people's future.

The evidence that the meteorite contains fossil bacteria amounts to the following. The meteorite contains hydrocarbon compounds (polycyclic aromatic hydrocarbons, PAHs), which are the same as the decay products of dead micro-organisms on Earth. It also contains amino acids, but these may be terrestrial contamination, which raises the possibility that the PAHs are similarly of Earthly origin. Countering this are indications that the isotopic composition of the material is not the same as is common on Earth.

The meteorite also contains mineral particles that, on Earth, are produced by bacteria. The particles are composed of magnetite. Magnetite particles in some species of terrestrial bacteria provide them with the ability to sense the magnetic field of the Earth, so they know 'up' from 'down'. The size and shape of the particles match the bacteria-produced magnetite we know, but not non-biological magnetite. There is no proof, however, that they are definitely of biological origin. It turned out to be possible to produce similar particles inorganically, under conditions that were plausibly similar to those that had been experienced by the rock during its time on Mars and later ejection into space.

Finally, and most dramatically, the meteorite contains tiny carbonate globules that are shaped like fossils of bacteria (32). The carbonate chemicals themselves are relatively common, and could have been introduced into the rock while it was on Mars or after falling to Earth. The shapes that the globules have formed into look as if they are fossils of biological creatures. The most striking was a tube made up of ten to twelve segments, tapering from a 'head' to a 'tail'. It *looked* like a bacterium. These fossils are small – they range in size from tens to thousands of nanometres.[3] Of course, some may be incomplete parts of a bigger creature; the rock has had a long, energetic history, and small fossils could well have been damaged. There has been some argument over whether the fossils that appear to be whole organisms can contain enough in the way of biochemistry truly to function as living organisms. But it has since emerged that they are at the extreme of the range of sizes of terrestrial bacteria (which can be as small as 100 nanometres in size) – but not completely outside the range.

It seems that what has been discovered in ALH84001 is provocative as an indication that there has been life on Mars, and we may have seen fossils of our first Martian. The evidence is the stronger for being clustered together in a small volume in cracks in the rock. But, considering each individual link in the chain of evidence, what appears to have been of bacterial origin could in each case have been made by something else non-biological. David McKay himself wrote, 'None of these observations is in itself conclusive for the existence of past life. Although there are alternative explanations for each of these phenomena taken individually, when they are considered collectively, particularly in view of their spatial association, we conclude that they are evidence for primitive life on Mars.' This is a strong claim, which the scientists whom McKay was trying to convince have said requires stronger evidence to substantiate it. To claim something so momentous as the existence of life on Mars, the proof must be convincingly conclusive. So far, the scientific community is not universally persuaded. If we can all agree on one thing, it should be that this episode shows how close biological activity is to other kinds of activity, and points to the difficulty we will have in recognizing life elsewhere – unless it is obviously alive, in the ways imagined by H. G. Wells. On the current evidence, however, extraterrestrial life will be what we make of it.

It seems impossible that Mars is populated by any beings that we could talk to. There are no artificial constructions, such as the once-hoped-for 'canals'. Nevertheless, Mars remains the best place to look for life outside the Earth. The Martian equivalent of the Hadean era ended at about the same time as ours, and it was bombarded by the same rain of asteroids and comets, bringing water and organic chemicals. Life got the same chance to start on Mars as on the Earth – it was, at its conception, undoubtedly a wet and (not quite so) warm 'little pond'. A global change in the planet's history stifled its development into a place where photosynthetic life could evolve and change its atmosphere from one of carbon dioxide to one of oxygen.

Did the climate change extinguish any life that may have started? As we saw in chapter 11, our own Earth underwent major transitions on more than one occasion, but life went on. The climate change on Mars was just this side of catastrophic. But, over the entire history of Mars, the Red Planet's climate has been modulated in cycles that range widely, because the Milankovitch cycles (see p. 137) that have affected Mars have been much more extreme than terrestrial ones. Mars has no large moon, and its rotation is not stabilized in the way that Earth's is (see p. 135). At times, over a period of thousands of years, Mars has been spinning with its axis almost perpendicular to its orbit round the Sun. There was not much difference between the climate in summer and in winter, and the ice in the polar regions was stable. But then, over a further period lasting thousands of years, the rotation axis of the planet progressively leant over, further and further. At times, Mars has been almost rolling along in its orbit, spinning with its axis near the orbital plane. As it went around the Sun, its polar axis would alternately point towards the Sun and away. The orbital period of Mars is approximately two years, so there would have been six months in which the north pole would have been directly in the light and warmth of the Sun, while the south was in darkness and unremitting cold, an intermediate period of six months during which the Sun passed from north to south over the Martian equator and warmed the planet more evenly as it rotated during its day/night cycle, six months in which the extreme seasonal differences of the north and south poles were reversed, and then another period of transition. During that time there were high-amplitude annual variations in climate across the entire planet. The rotational axis of the Earth is currently tilted at an angle of 23.5° and has oscillated over a range of 3°; the rotational axis of Mars is currently tilted at a similar angle of 25.2° but has oscillated over a range of 20°. So the climate of Mars has changed very much during its history, even disregarding the global catastrophe of its loss of atmosphere. Life on Mars – if it exists – has repeatedly faced the challenge of very tough, highly variable conditions, but, having begun – if it did – and developed over the history of the planet, it may still endure in the present climatic conditions – surviving, struggling and little-developed, in caves or chasms under the surface.

14

Ice World: Europa

In 1609, Galileo Galilei (1564–1642), professor of mathematics at Padua University in the Venetian republic, heard about the invention of the telescope by a Dutch optician the year before. He made a prototype of a telescope that he promoted (profitably) to the Venetian authorities to assist them in their maritime activities. At the end of 1609, Galileo began to use the telescope for celestial observations, looking up at the Moon and the stars. In the first days of 1610 he observed the planet Jupiter for the first time. In a letter of 7 January 1610, Galileo wrote about a momentous discovery that he had made the previous night:

> And besides my observations of the Moon, I have observed the following in other stars. First that many fixed stars are seen with the telescope, which are not otherwise discerned; and only this evening, I have seen Jupiter accompanied by three fixed stars, totally invisible by their smallness...

At first Galileo did not think there was anything remarkable about these stars – a trio in a straight line through Jupiter, two on one side and one on the other – except that they existed, and could not be seen without a telescope. According to his observation journal, when he came to look at Jupiter again on 8 January there were still three stars, but all on the other side of the planet. Presumably, thought Galileo, Jupiter had moved on from its position amid the three. On the 10th and 11th there were two stars only, on one side, with the third (Galileo speculated) conjoined with Jupiter. On the 12th the three stars were two on one side, one on the other once more – he noted, 'It appears that around Jupiter there are three moving stars invisible to everyone up to this time.' On the 13th, for the first time, Galileo saw that there were actually four little stars.

It seems that Galileo thought at first that the stars moved back and forth in a straight line. But if this was the case, how did they pass through Jupiter and through one another? Suddenly Galileo realized that the four 'stars' were satellites in orbit around Jupiter. Their orbits were edge-on to the line of sight to Earth and although they described a circle around Jupiter, their motion appeared as a straight line by projection.

When Galileo wrote up his discoveries, in a book called *The Starry Messenger* (published in 1610), he dedicated it to Cosimo II de' Medici (1590–1621), and proposed to call the satellites of Jupiter 'the Medicean stars'. But the names that stuck were those first mentioned by the astronomer Johannes Kepler, then working in Prague, and promoted by Simon Marius (1573–1624) in 1614: Io, Europa, Callisto and Ganymede, all of them in mythology Jupiter's lovers.

When Galileo looked in his telescope at Europa, the second-closest satellite of Jupiter, all that his eyes could see was a featureless point of light. In his mind, he imagined another world, such as our Moon, but he could not then conceive of how strange a place Europa would turn out to be: a waterworld, its surface frozen over with deeply cracked ice (see p. 65). Europa is the archetype of a particular kind of habitable environment – an ocean of water within an icy world – that may be the most abundant habitable environment in the Universe.

Galileo refrained from speculation about the nature of Jupiter's satellites, and whether any of them sustained life. But Johannes Kepler was less inhibited on the question. Kepler is best known as the man who discovered the laws of planetary motion, which were the inspiration for Isaac Newton's theory of gravitation. But he was also an astrologer and a mystic. He used the existence of Jupiter's moons as a proof that Jupiter was inhabited. In 1611 he wrote about *The Starry Messenger*: 'where the story is told of the discovery, by the aid of a very good telescope, of another world similar to ours in the planet Jupiter'. The four satellites must have been made not for us, but for inhabitants of Jupiter, because up to then they had never been seen by anyone on Earth. God made the moons, and He does everything for a reason, so there must be people on Jupiter who can see them. 'The conclusion is quite clear,' Kepler went on. 'Our Moon exists for us on the Earth, not the other globes. Those four little moons exist for Jupiter, not for us. Each planet in its turn, together with its occupants, is served by its own satellites. From this line of reasoning we deduce with the highest degree of probability that Jupiter is inhabited.'

In one respect we find Kepler's argument absurd, since science no longer accepts that celestial objects are made for us or for anyone else. But in another way his viewpoint was profound for its time, since it put the Earth on a par with Jupiter, and human beings as equals with the presumed inhabitants of other worlds.

* * *

We now know that the satellites that Galileo discovered form a family of four similar-sized spherical worlds, 3,000–5,000 km in diameter, the four largest of about sixty moons that orbit Jupiter. With terrestrial telescopes, it is difficult to study these small worlds at such a great distance; their individual characteristics were first discovered by the two Voyager spacecraft that visited Jupiter in 1979. A lot more detail about the moons has been gathered through prolonged examination by the Galileo spacecraft, which was in orbit around the planet from 1995 to 2003. The moons of Jupiter are surprisingly different from one another.

Io is the closest of the larger satellites to Jupiter. In 1979 Linda Morabito (b. 1953), a young engineer, discovered from images in Voyager's navigation cameras that Io has active volcanoes, which eject ash high above the satellite's surface. Its surface is covered with volcanic craters, lava flows, and drifts of ash. Some of the lava flows are solidified, but some are also – from time to time – liquid and flowing, red hot, down the volcano slopes. Io will have had impact craters in the past, but they have all been covered up by these volcanic deposits. It is hard to imagine that life would have evolved or taken a hold on this dry, fiery world, with its surface constantly renewed, undermined by volcanic eruptions and strewn from above with lava and ash.

The outer two of Jupiter's satellites, Callisto and Ganymede, are rocky bodies, held together with ice, covered with impact craters, like the Moon, and with bright, white areas of pure water ice on their surfaces. Ganymede is the largest satellite in the Solar System; it is bigger than the planet Mercury. Callisto is the size of our own Moon, and has been similarly geologically inactive for nearly all its life, with a surface covered with craters expressing the bombardment history of the entire 4.5 billion years of the age of the Solar System. By contrast, the exterior of Ganymede, which also has many craters, was only geologically active for perhaps the first billion years of its past. The surfaces of both are now frozen, with temperatures averaging about –130°C, but there is some evidence from the behaviour of the magnetic field of Ganymede – and the way that it disturbs the magnetic field of Jupiter – that a salty ocean of water lies some 200 km below its surface. Think of a metal detector at an airport. As you walk through, anything that conducts electricity – metal jewelry or weapons especially, but also the water in your body – disturbs the magnetic field. It is the same with Ganymede. The water that it contains is maintained as a liquid by heat from radioactivity or some other source within the satellite. Some areas of the moon appear to be dry salt lakes, coated with minerals left behind by the evaporation of salty water on the surface. They are possibly the result of brine making its way to the surface by eruptions or through cracks. (Common salt, the chemical sodium chloride, NaCl, which is dissolved in seawater, is the result of weathering by rain on the surface rocks of the Earth, and there is no such weather on such satellites as Callisto and Ganymede. The salts on the surface of Ganymede are general mineral salts, not necessarily sodium chloride.)

* * *

The second satellite outwards from Jupiter is Europa, which has the distinction of being one of the most spherical objects in the Universe. Europa is about the size of Earth's Moon – it is also basically a rocky, icy body, but it is warmer than Callisto and Ganymede. It too shows disturbances as it passes through Jupiter's magnetic field, and betrays the presence of substantial amounts of water in the satellite. The ice that permeated its interior has melted, risen to the surface and formed an ocean. The ocean is frozen at the top, so Europa is covered with icy plains, criss-crossed by grooves. The grooves are cracks in the ice, and the plains are in fact a jigsaw of ice floes the size of cities (**33**). Frozen 'puddles' smooth over older cracks.

Water has seeped up through the cracks from the ocean below, and stained the surface of the ice floes with coloured salts. There are very few meteor craters on Europa's surface – when one does impact on the ice, the meteorite and the chemicals that it brings with it sink into the ocean, and the shifting ice floes close up overhead and quickly erase the traces of the crash. It is estimated that the surface renews itself totally in a time of about 50 million years. The ice layer of Europa is perhaps a kilometre or more thick, and it is the pressure of the floating ice and radioactive and tidal heating of Europa's interior that liquidizes the water. The ocean may be as deep as 100 km; the average depth of Earth's oceans is only 5 km. Europa has more water than the total amount found on Earth, a salty ocean under an icy cover. It seems possible that life has evolved there.

The reason why Europa and Io are so different from the cold, rocky satellites Callisto and Ganymede is that they are closer to Jupiter, and experience its strong tidal force. They are compressed and released, pumping like a heartbeat. The energy of the movement releases heat, and raises their internal temperature. The abnormally high temperatures inside Io fluidize its rock, which bursts from the surface in volcanic eruptions. In the case of Europa, the volcanic eruptions are on the ocean floor. In other words, Europa probably has some form of 'black smokers' (see p. 93). The temperatures inside this moon, in conjunction with the high pressure under the ice, keep its ocean as liquid water. The salts that show on the surface of the ice imply that black smokers have pumped Europa's ocean full of dissolved minerals, and it is briny.

* * *

Because ice clearly exists on Europa, and it has a briny ocean similar to Earth's, as well as sources of volcanic heat and minerals that could fuel life, many scientists think that living organisms could have evolved there. They point to similar habitats on Earth where life is abundant, particularly in Antarctica, where every kind of creature is represented in the water below the ice. At the shoreline of the continent (underneath the ice) live sponges, crinoids, scallops, snails, fish and many kinds of micro-organisms. Even the lakes of Antarctica, some of them frozen and isolated for possibly millions of years, contain life.

One of the most extreme of these lakes is Lake Vostok, under the Russian Vostok Research Station on the central Antarctic plateau at an altitude of 3,500 m. There had been indications of the existence of this body of water since the 1970s, from airborne and space-based radar observations; it was delineated, one could even say discovered, in 1996. It is 250 km by 50 km in area, and its surface is below sea level and up to 800 m deep. In size it is comparable to one of the Great Lakes on the Canadian–US border. It may have been sealed off from the outside world for up to 15 million years. No samples have yet been taken of the liquid from the lake itself, and when the breakthrough occurs (under carefully controlled conditions, so that nothing drops in from the surface to contaminate the water) biologists may find nothing living at all; they may find an ecology similar to that on Earth now; or they may find some living fossils of creatures that have evolved independently for millions of years, from organisms of the distant past. By 2010, ice cores had been drilled to depths within 100 m of Lake Vostok's surface, and as I write, the breakthrough is imminent. The samples brought up from the depths so far contain living and latent micro-organisms similar to those found on the surface. Comparable results have been obtained from ice cores drilled into the permafrost in Greenland and Siberia.

These Earth-based results test the lower limit of temperature at which archaea and bacteria can survive. We know of bacteria that are active at temperatures down to –20 and –30°C, but some of the essential chemical reactions operate at as low as –120°C. At the other extreme, bacteria can be active at up to +120°C. Other limits are set by the range of pressure, acidity and salinity (saltiness) over which bacteria and other simple life-forms can survive. The ones that push the boundaries are called 'extremophiles' (lovers of extreme conditions). The various terms for different sorts of extremophile are given in **Table 7**. People who search for extremophiles[1] have found them not only in Antarctica but also in the hot geysers of Yellowstone National Park, in the vicinity of black smokers in the Atlantic trenches, at extreme oceanic depths, below ground up to 3 km, in the Great Salt Lake of Utah, in the Atacama Desert and so on. There may be extremophiles that push out even beyond the conditions survived by these organisms, if there is the evolutionary imperative to do so.

Table 7. Extremophiles

	An organism that survives or thrives...
Acidophile	in extreme acidity.
Alkaliphile	in extreme alkalinity.
Barophile	at extreme pressures (see also piezophiles).
Cryophile	in extreme cold.
Cryptoendolith	in microscopic spaces within rocks.
Endolith	deep within the ground.
Halophile	in very salty water.
Hyperthermophile	at high temperatures in hydrothermal systems.
Hypolith	underneath rocks in the desert.
Lithoautotroph	by consuming carbon from carbon dioxide and deriving energy from minerals.
Metallotolerant	in water in which there are high concentrations of heavy metals.
Oligotroph	on very little nutrition.
Osmophile	where there is a lot of sugar.
Piezophile	at high pressure.
Polyextremophile	in multiple extreme conditions.
Radioresistant	under strong radiation (ultraviolet or nuclear radiation).
Xerophile	where it is very dry.

The sea on Europa is on average cooler than −60°C, which is beyond the known limits at which bacteria or archaea can be active. But this average may conceal variations in temperature associated with geothermal hot spots, or perhaps the micro-organisms on Europa are especially efficient.... Many factors could justify life on Europa, but thus far all are unproven possibilities.

It seems clear that it is worth investigating Europa as an abode for extra-terrestrial life. Such life would be aquatic, and able to survive cold. Given the ice that covers the sea, the life-forms would not have to be radiation resistant. On the other hand, this covering will make it dark in the oceanic depths; anything resembling photosynthesis, relying on the diminished sunlight and the light reflected from Jupiter, could take place only in the cracks in the ice that reach up to the surface. The individual creatures could be surprisingly large, since mechanisms have been found that transport oxygen made by the effects of radiation at the surface of Europa into the sea through the cracks between ice floes – if life evolves to use oxygen generated in this way, it can be more efficient and grow larger, bypassing the need for there to be an oxygen atmosphere generated by photosynthesis. We may therefore hope to find fish in Europa's ocean. There has certainly been time for complex creatures to evolve – this icy moon has been stable for billions of years. In fact, Europa, and others like it, may be the most common habitat in the Universe for complex extra-terrestrial life.

It will not be easy to look into Europa's ocean. Any spacecraft that is sent will have to land by reverse-thruster rockets, since this moon has no atmosphere. If the craft lands safely it will then have to penetrate a thick ice cover. It is likely that a drill would be impractical, so attention has turned to methods of inserting a probe through the ice by melting the latter using nuclear energy. Maybe this is too difficult for the first craft that visits Europa, and it may be enough to send a lander to take a close-up look at the surface.

It is possible that the lander itself may be able to find life. Just as the ice of Antarctica churns and excavates the lower ice and thus drives its contents (such as meteorites that had fallen long ago) to the surface, so also do processes on Europa – we see the results of this churning as coloured stains on ridges of ice at the boundaries of its ice floes. Perhaps in these coloured stains lie dead creatures, brought up from the depths of the ocean and exposed to view by orbiting spacecraft or landers that can rove over the surface.

15

Glimpses of the Old Earth:
Enceladus and Titan

Galileo saw Saturn through his telescope in 1610, and observed something strange either side of the planet: two ear-like extensions. He thought that these might be two large moons. Over the decades, the appearance of the planet changed and the 'moons' disappeared. We now know that Saturn is surrounded by thin rings that all but vanish when the Earth passes across their plane. It was in 1659 that the Dutch physicist Christiaan Huygens, with his greatly improved telescope, was able to see the form of the rings. They are made up of innumerable small bodies – individual small rocks and chunks of ice, ranging in size from boulders to gravel – in orbit around the planet. The most common theory of their origin is that these are the pieces of satellites that strayed too close to Saturn and broke up under the tidal force of the planet.

Apart from those bodies in the rings, Saturn has more than sixty moons, but only a dozen are of any size (greater than 50 km). They were discovered in groups, starting with the largest and progressing, as telescopes increased in capability, towards the smallest. The first to be discovered, by Huygens, was indeed the greatest in size. Named Titan, it is more than 5,000 km in diameter: bigger than the Moon, or even the planet Mercury. The sixth and seventh largest are called Enceladus and Mimas, and were discovered by the British astronomer William Herschel in 1789. Enceladus is 500 km in diameter, Mimas 400 km; both are comparable in area to such a country as Britain, and are on the small side when considered as worlds.

* * *

Saturn was first explored by spacecraft in brief flybys by Pioneer 11 in 1979, and by the two Voyager spacecraft in 1980–81. The Saturn system is currently being studied by the NASA/ESA Cassini spacecraft, one of the most ambitious planetary exploration missions ever. Probably 4,000 people worked on this spacecraft, and the mission cost about $3 billion. The spacecraft Cassini was launched in 1997 from Cape Kennedy in Florida. I stood beside a channel draining the marshland of the peninsula where the spaceport is located, and watched the night-time launch. At my feet (and two or three metres down a vertical bank!) an alligator stirred in the greasy water, moonlight sparkling from its lazily blinking eyes and glistening scales, a living fossil from the era of the dinosaurs. A few kilometres away two tonnes of the most sophisticated late-twentieth-century technology sat on top of a Titan IVB/Centaur rocket, waiting to send the spacecraft to places where, if there was life, it would likely be far more primitive compared to the alligator than the alligator was compared to me.

The Titan rocket had been selected for the launch, not because it was appropriately named to start the exploration of Saturn's largest satellite, but because it was the most powerful rocket then available to space scientists, and its power was necessary to push such a massive spacecraft such a long distance. The rocket lifted off into the night, its bright exhaust gases blazing through a thin layer of cloud. Its smoky trail curved away from where I stood, over the Atlantic Ocean, as the rocket picked up speed, using every possible advantage of the Earth's rotation. There was a puff in the smoke trail as the rocket jettisoned its first stages and its fiery exhaust faded from sight as the spacecraft left the Earth. For seven years the spacecraft toured the Solar System in a circuitous journey to Saturn – twice round Venus, then past the Earth, then Jupiter – picking up speed at each flyby. Cassini arrived at Saturn in 2004 and began its programme of exploration, looping repeatedly through the Saturnian system at the behest of its controllers. It is expected to stay in orbit there until 2017, when it will plunge into Saturn's atmosphere and burn up.

These spacecraft established that Saturn has Jupiter-like clouds and weather, though it is not as active as Jupiter, due to Saturn's greater distance from the warmth of the Sun. The space cameras measured the rings and observed their complex, braided motions. They mapped satellites, including Mimas – remarkable for its gigantic crater, like the eye of a Cyclops or the communication dish on Darth Vader's Death Star. The crater was caused by the impact of an asteroid of such a size that it must have come close to destroying Mimas altogether – indeed the shock of the impact wrecked the surface and created grooves and chasms that cover the moon.

*   *   *

The satellite Enceladus is also a small, rocky and icy sphere, but its surface is remarkably varied. The terrain that covers its north pole is old and heavily cratered, like our Moon. The craters, however, are distorted, eroded and cut by chasms. There has evidently been considerable geological activity since they were formed. This perception is confirmed by the smoother, wrinkled terrain of the southern hemisphere of Enceladus. It is cut by large dark cracks, called 'tiger stripes', which are very young, and surrounded by ice that has welled up from below (34). Judging by the paucity of meteor craters, the smoother terrain is less than 500 million years old.

Evidently Enceladus has been recently active. In fact, it is active even now. Cassini recently discovered that there are sprays of water and water-ice chips (hail and snow), in a gas of methane, carbon dioxide and other simple organic molecules, bursting from several vents – probably from the 'tiger stripes' – in the south polar region. The existence of the sprays had been suspected from flybys over Enceladus, when the gases were directly detected. The sprays were imaged on another pass over the surface in 2006, when Cassini was positioned specifically to view the sprays, backlit by the Sun (35).

It seems that the sprays of snow and hail last for millennia, possibly for millions of years. The snow falls back onto the surface of Enceladus in a repeatable pattern, and areas of the surface have been blanketed in a thick layer of tiny ice particles. The resulting terrain is unusually smooth, with ghost-like undulations indicating buried canyons and craters. The biggest of these canyons are 500 m deep and 1.5 km across, not unlike those found in the American Southwest. The overlying layer of fine, powdered snow is sometimes 100 m deep in this area, having accumulated at a rate that is extremely slow by terrestrial standards – less than a thousandth of a millimetre per year – but which has built up a fine piste for skiing.

The sprays of ice that have produced this potential year-round playground for winter sports are geysers fed by reservoirs of liquid water perhaps not far below the surface of the south pole. The source of the heat that drives these geysers is not known – the flexing of the body of the moon by the tidal forces of Saturn, in the same way that Io and Europa are flexed by Jupiter, is not energetic enough. But however it happens, inside Enceladus is hot rock, the warmth of which melts its internal ice and fills its underground caverns with water, laced with organic chemicals in solution.

This environment is not unlike some niche habitats on Earth, which are able to maintain established ecosystems. In the dark, deep within volcanic rocks, the food chain begins with bacteria that consume hydrogen or sulphur produced by reactions of the heated rock with water.

The implication is that Enceladus is a potential habitat for life. If Darwin was writing today about where life originated, he might, instead of the phrase 'warm, little pond', refer to a 'warm, giant cistern'. According to Carolyn Porco (b. 1953), head of the imaging team for the Cassini mission, 'This discovery brought Enceladus up to the forefront as a major target of astrobiological interest.' The geysers bring samples of the 'warm pond' above the surface of the moon, where they could be collected for analysis by a spacecraft searching for extraterrestrial life; it would not even have to land, let alone bore through a kilometre of ice.

* * *

The first flybys of the space programme to explore Saturn discovered something unique about the planet's largest moon, Titan. It appeared to have been a blank sphere, with no visible surface features. Above its limb (outer spherical edge) hovered an orange haze. Titan is the only satellite in the Solar System with a substantial atmosphere. It extends nearly 1,000 km into space. It is composed mostly of nitrogen, but a few per cent (1 to 5 per cent, depending on atmospheric height) is methane, with traces of such noble gases as helium and argon, and other hydrocarbons. The methane is the source of the haze, a hydrocarbon smog of sooty particles produced by the action of the Sun's ultraviolet light on the atmosphere. The continued existence of the atmosphere of Titan is perplexing, since in theory the action of sunlight should convert the whole atmosphere to other hydrocarbons within 50 million years. There must be a source of methane within Titan itself, but it is not known what it is. Perhaps there are reservoirs of the gas below the surface, released through volcanic vents; but no such events have been unambiguously identified. Some people even speculate that there might be biological activity producing the methane.

The surface temperature of Titan is –170°C, and the surface pressure is one-and-a-half times that at the surface of the Earth. At such a low temperature, methane condenses to a liquid. So Titan must have a 'wet' surface, saturated not with water but with liquid methane. The nature of the surface was unknown until the Cassini mission – was it watery everywhere, with seas and lakes? Splattered with puddles among rocks? Soggy like a marsh?

The Cassini mission was in fact two spacecraft. The second was a lander, called Huygens, that was pinned onto Cassini and carried piggy-back to Saturn. Huygens was allowed to 'sleep' for the seven-year flight, being woken briefly every six months for a health check. On arrival at Christmas 2004, Cassini released the Huygens lander to fall and then parachute onto Titan two weeks later in an attempt to establish the moon's nature. The lander had been powered with batteries that lasted for three hours, of which two-and-a-half hours powered the slow descent through Titan's atmosphere, during which time the spacecraft measured the atmosphere's composition and other characteristics. Swinging like a pendulum below the parachutes, drifting in the wind, the spacecraft was entirely autonomous during its descent, for the simple reason that by the time radio waves notifying controllers of trouble and asking for help had travelled to Earth and got an answer, three hours would have passed and the problem would have been resolved, one way or another.

A downward-facing camera established that the spacecraft was headed towards a rocky shoreline that divided a flat plain from hills riven by drainage channels: rivers of liquid methane. On which side of the shoreline would it land? Extending below the lander was a thin probe, the first terrestrial object to touch Titan, designed to see how hard the landing-place was – would the lander hit the surface with a splash, a bang or a squelch?

If anyone had been listening, they would have heard the Huygens lander hit the ground with a soft thump. Huygens touched down on a relatively smooth, but not completely flat, surface that was neither hard (like solid ice) nor very compressible (like a blanket of fluffy aerosol); rather, the craft landed on a relatively soft, solid surface, with a texture similar to lightly packed snow or sand. The lander settled gradually, by a few millimetres, after arrival, as its weight compressed the pebbles on which it had landed into the sand. The area around the landing-site showed rounded boulders, tumbled down the river channels onto what must have been a lakebed, flooded at times of prolonged methane rain. Later in the mission, using radar techniques, the Cassini spacecraft imaged methane lakes (36) at one of the poles of Titan, high waves rolling slowly across their surface, whipped up by the high winds.[1]

The atmosphere of Titan resembles that of the Earth in former times, and Titan's rich carbon chemistry is thought to resemble the conditions that prefigured life on Earth. In experiments at laboratories in Paris and Grenoble carried out in 2009, modelled on Stanley Miller's tests (chapter 7), but starting with an atmosphere akin to Titan's and using radio energy as the input (in analogy with sunlight), Sarah Hörst, a PhD student at the University of Arizona supervised by Roger Yelle, found that the mixture produced the five nucleotide bases used by life on Earth (cytosine, adenine, thymine, guanine and uracil) and the two smallest amino acids, glycine and alanine. The chemical ingredients for life can be made in Titan's prebiotic atmosphere. Stanley Miller's experiments were intended to simulate processes taking place on Earth, in oceans and rainstorms, but Titan has no liquid water. Hörst's experiment used no liquid water in the chemical mixture that she activated. It was the first time that molecules such as nucleotide bases and amino acids had been found in such an experiment, starting without liquid water. If life, some sort of archaea perhaps, has evolved on Titan, then it has not, for some reason, taken the moon's environment to the stage of the Great Oxygenation Event that altered Earth's atmosphere about 2,000 million years ago. It will be intriguing to explore Titan further to see if archaea can be identified in the methane lakes. If so, it will be the evidence for which the space-science explorers have been searching – a link with the past that goes right back to the end of the Hadean eon on Earth, and the start of life here.

Together, Titan and Enceladus may hold the answer to the question, 'Where exactly on Earth did life start? Was it in the volcanic ocean depths, or a choking, organically laced atmosphere?'

16

Are We Lonely, or Really Alone?

When Carl Sagan estimated that there are millions of intelligent civilizations in the Galaxy (see chapter 3), he may have been over-optimistic. Under the name 'Rare Earth', some modern theories have identified lucky flukes in the history of our Earth that have given our planet properties uniquely favourable for intelligent life, such as the stability necessary to give it time to develop. The name is derived from a book, *Rare Earth: Why Complex Life Is Uncommon in the Universe*, by Peter Ward and Donald E. Brownlee (2000). The authors produce arguments that suggest that all of the factors in the Drake Equation multiply up to $N = 1$; there is only one planet in our Galaxy with intelligent life, namely the one we live on.

The factors in the Drake Equation are described on pages 31–37. R_*, the birth rate of stars in our Galaxy, and the fraction (f_p) that have planets, are relatively non-controversial calculations. The number (n_e) of earths per planetary system is not as clear cut. Our Solar System has four terrestrial planets, but in the planetary systems discovered so far, the number is less. In fact, in the sample of thousands of exoplanets now in the catalogues, astronomers have detected just a few planets that are possible earths. Why is this? Is it because earths are very rare? Or because astronomers are incapable of finding any? It is the latter: the way that planets are found favours finding the larger ones. The way that the sample of exoplanets has been selected leads to a bias towards the conclusion that earths are rare, because earths are too small for our currently available technology to find them readily. They are not very massive and do not swing their parent star around much, and if they pass in front of their parent star they will block off a small fraction of its light, such a small fraction that it is hard to detect. We cannot rely completely on the apparent statistics in concluding that the number of earths per planetary system is few – absence of evidence is not the same as evidence of absence.

But there is some indication that earths are not the routine presence in planetary systems that we have hitherto believed, on the basis of the single example of our own Solar System. Most planetary systems contain so-called 'hot jupiters': gas giant planets (see p. 56) that were formed beyond the 'snowline' in the outer part of their planetary system, but that have migrated inwards to arrive at a distance close to their parent star. Hot jupiters are not just theoretical – with the Hubble Space Telescope, astronomers have seen signs of such planets being evaporated by the heat of their sun. Hot jupiters are relevant to the question of the number of earth-like planets in a planetary system, because they compete with earths to live in the inner regions of the system. Earths are born in the hot inner parts of their planetary system out of material that has been warmed so that its ice and gas content has evaporated. Jupiters are gas giants, born in the cold outer regions of a planetary system where the ice and gas can survive. So if a jupiter is now hot, it must have migrated inwards from its birth-place. Migrate a massive planet into the same area of a planetary system as an earth and the earth will be disturbed in its orbit. It may plunge into its sun. It may be thrown into space. Or it may be pulled into the jupiter and cannibalized.

The reason why jupiters migrate inwards is that they interact with the planetary disc from which they have formed. The massive planet causes waves and concentrated ridges in the disc, which represent concentrations of mass that pull on the jupiter. The competing pulls from the inside and the outside of the jupiter's orbit are not equal and do not cancel each other out. There is a net pull inwards.

Evidently our own Jupiter remains in the outer reaches of the Solar System. It may have been pulled inwards during the formation of the Solar System when there was a nebular disc, but the pull inwards was turned off at some point – Jupiter itself was formed further out of the Solar System than we now find it, and it may have migrated inwards for a time, but it stopped drifting in towards the Sun. Perhaps the material of the disc became exhausted as it condensed into the planets and there was only a very thin disc remaining to do the pulling, which had little effect before it all dissipated. Jupiter never rampaged through the inner Solar System and therefore never wreaked havoc on its terrestrial planets. Lucky for us.

What we do not know is how common the migration phenomenon is, how often it destroys inner earths, or how typically it stops before doing so. We know only that migration is common in the planetary systems discovered. But since those systems are likely to be the ones with massive jupiters orbiting close to their parent star, there is a strong bias in the examples that we have right now, another selection effect in the statistics; we are still in the dark on this question.

It is not inevitable, during the migration of a jupiter through the inner protoplanetary disc, that any earths are destroyed. According to some simulations, half the material in the inner disc might be thrown outwards, including some, but not all, of the embryonic earth-like planets. Nevertheless, migration certainly reduces the number of earths in a planetary system.

A second uncertainty in the Drake Equation is the probability that life on an Earth-like planet develops complexity and intelligence. What is required for this, judging by the only example we have – our Earth – seems to be time (see chapter 9) and stability – at least, stability within some bounds of variability. An erratic orbit, which causes a planet's climate to oscillate to extremes, might be a problem, as would an unstable liquid-iron core, which would cause the planet's magnetic field to collapse, as is the case of Mars. If a significant part of the reason why Earth has a large core and a stable orientation is because a chance collision produced our Moon, then for this long-term stability to replicate elsewhere could be very rare indeed. Asteroid and comet impacts, and their effect on an earth's climate, could also be a problem. As we have seen, asteroids and comets were largely cleared away at the time of the Late Heavy Bombardment early on in the history of our Solar System, however, leaving smaller and fewer asteroids to hit the Earth while life was evolving; so this favoured the development of life on Earth in a way that also might not be replicated in other planetary systems.

Finally there is the length of time, L, during which a planet supports intelligent radio-transmitting life. There is a lower limit from data about the Earth – there has been life on this planet transmitting strong enough radio signals for fifty to ninety years. If our society or even species becomes extinct in the near future, after nuclear war, anthropogenic global climate change, exhaustion of such natural resources as oil or minerals, having too many children and starving each other to death, or some other dystopia, and this outcome is typical of intelligent civilizations, then, because the lifetime of radio-transmitting civilizations is therefore small (less than a hundred years), the number in the Galaxy at any time, found by multiplying all the different factors in Drake's Equation, is small, about $N = 1$, and at the moment we are it.

How likely is this fate for us in the immediate future? It is certainly true that since about 1945 and the invention of nuclear weapons we have had the power to affect the planet on a global scale. It was in 1947 that the *Bulletin of the Atomic Scientists* started annually to depict how close we are to a world disaster through its Doomsday Clock, showing how far we are from midnight closure. It was originally set at seven minutes to midnight and has been as close as two minutes to midnight at the height of the Cold War between the USA and the USSR. In 1991, at the time of the Strategic Arms Limitation Treaty, it was at its furthest from midnight, showing seventeen minutes to spare.

Since 2007 the Doomsday Clock has taken into account other threats to human-ity, and crept back to five or six minutes to midnight, based on global climate change and the proliferation of nuclear weapons in the hands of unstable states. Unfortunately the Doomsday Clock is not calibrated. It is used as a designer's symbol to indicate graphically the changes in the risks to the world as the hands get closer to or further from midnight, not as a literal estimate of time left. It is unclear as to what 'five minutes' really means, but some gloomy predictions suggest a period of about a hundred years is left to us.

It seems that the originally high numbers offered by the Drake Equation for the number of civilizations in the Galaxy could be optimistic, for a number of different reasons. The answer may be that there is just one; this is known as the Rare Earth hypothesis.

*   *   *

The Rare Earth hypothesis is a solution to what has become known as the Fermi Paradox. The story is that, in 1950, American newspapers had been following two threads of extraordinary stories. First, public trash cans had mysteriously been going missing. Secondly, there were many reports about flying saucers, piloted (perhaps) by extraterrestrial aliens. The noted physicist Enrico Fermi (1901–1954) was at the time working at the Los Alamos research facility, and one lunchtime he and some colleagues saw a cartoon that linked the two stories. A group of extraterrestrial aliens were shown having returned to their home planet, unloading trash cans from their flying saucer. This pro-voked the group of scientists into a discussion of interstellar travel and the possibility that Earth was being visited by extraterrestrial aliens. Suddenly Fermi put the question 'Where are they?'[1]

What Fermi had in mind was that if intelligent life was ubiquitous in the Galaxy then there would no doubt be some less and some more advanced than us. Technology has developed very rapidly; here on Earth, as Fermi knew at that time, we have gone from no flight at all to supersonic travel within fifty years, and as we have seen in the last half-century, the exploration of space will happen in less than a further fifty. The Russian astrophysicist Nikolai Kardashev (b. 1932) imagined three levels of civilization. Type I civilizations control the energy of a planet, Type II the energy of a star, Type III the energy of a galaxy. Humans on Earth are already nearly at the level of Type I. So if there really are millions of civilizations in the Galaxy, we might expect some of them to be at a level of Type II or more. It seems that some of the more advanced extraterrestrial aliens would have developed interstellar transportation methods and ought to be with us here on Earth now. They are not.[2]

Perhaps interstellar travel is impossible. Perhaps it needs too huge a leap in technology. Perhaps to travel interstellar distances in a feasible time requires scientific factors, such as faster-than-light speeds, that are not only impossible at the present time, but will also remain impossible for ever. Perhaps it needs energy sources that can be quite easily numbered Type II in a theoretical exercise but which do not exist. Perhaps interstellar travel is inevitably fatal, and people will be unwilling to participate in it as astronauts. But this does not explain why we have detected no radio signals. The science writer David Brin (b. 1950) called this 'the Great Silence' in a paper published in 1983 by the Royal Astronomical Society. He concluded that:

> The quandary of the Great Silence gives the infant study of xenology [the study of extraterrestrial intelligence] its first traumatic struggle, between those who seek optimistic excuses for the apparent absence of sentient neighbours, and those who enthusiastically accept the silence as evidence for humanity's isolation in an open frontier.... Some of the branch lines discussed here serve the optimists while some seem pessimistic to an unprecedented degree.

Brin's own optimistic conclusion is that 'It might turn out that the Great Silence is that of a child's nursery wherein adults speak softly, lest they disturb the infant's extravagant and colourful time of dreaming.' In other words, alien intelligence exists and has chosen not to visit overtly or transmit messages to us because we are immature and will not be able to cope with the messages. Astrophysicist Paul Davies (b. 1946) is more pessimistic. His term for the failure to discover radio transmissions from intelligent civilizations is the Eerie Silence. He thinks that the life on Earth may have been a one-off fluke, and that extraterrestrial intelligence does not exist.

Other astronomers – David Darling (b. 1953), for example – criticize the Rare Earth approach. In his book *Life Everywhere: The Maverick Science of Astrobiology* (2001), Darling writes:

> What matters is not whether there's anything unusual about the Earth; there's going to be something idiosyncratic about every planet in space. What matters is whether any of Earth's circumstances are not only unusual but also essential for complex life. So far we've seen nothing to suggest there is [something so unusual that complex life could not evolve elsewhere].

If the Rare Earth theories are right, life may be ubiquitous in the Universe but intelligent life may be rare. Two thousand years of astronomical endeavour have not yet proved conclusive enough to decide either way.

* * *

Science has established beyond doubt that the Universe is swimming with other planets. It seems likely that some of them are Earth-like (although we have not found any particularly close matches yet) and that others have niche environments where life is possible (on Mars, and on some of the other planets and moons of the Solar System). Science has added the distinct possibility – even probability – that life exists in some of these places in the form of archaea or bacteria. There is every chance that we will, within the lifetime of many readers of this book, find such creatures in our own Solar System by travelling to well-targeted planets with robotic or manned spacecraft.

On the question of whether we might find intelligent life elsewhere in the Cosmos, the science has advanced but the same fundamental uncertainty remains. Our planet has survived the vicissitudes of our cosmic environment and produced us. Other planets may not be so stable or so fertile – or so lucky. Life in some elementary form may be common in the Universe, but complex and therefore intelligent life may be rather rare. If so, there will be large distances, measured perhaps in thousands of light years, between planets that have intelligent life. This will make it difficult if not impossible to communicate with the aliens that live there, and has certainly prevented us from getting in touch so far. We may, in our hearts, like to take comfort in the thought that life is everywhere, but we must also face the rational possibility that to all practical effect we are doomed to be lonely, if not exactly alone. If the most likely outcome of the search for extraterrestrial life in the short or medium term is that we find something similar to a bacterium or archaeon, or another being that differs from us, it will not provide much psychological comfort, although the technological and scientific rewards could be significant.

In a report in 2006 on the science of the search for extraterrestrial life, NASA took an empirical view:

> To understand whether life is common or rare in the universe we
> must meet the technological challenges and embark on a series of
> complementary and interlocking explorations, each activity
> supporting and extending our efforts to characterize planetary
> systems and search for terrestrial planets and life.

I changed my mind as I wrote this book. I began as the Greek atomists did, thinking on general grounds that it was impossible for us to be the only intelligent life in the Cosmos. I was rather surprised to find, as I worked on it, that it is conceivable that intelligent life on Earth is, to all intents and purposes, unique. It is true that the practical search for life started only recently, and we have not been able to look far enough afield to conclude that there cannot be intelligent life elsewhere. But the only fact that we know for sure about such life is that it exists here on Earth, in a rather limited zone that we dignify with the name of biosphere; and it exists here in rather specific cosmic circumstances. If there is complex life elsewhere, it would seem that it is very rare, quite likely not as advanced as humans (possibly hardly advanced at all), and probably a very long way away indeed. Or possibly, since life on Earth will be extinguished in the near future by some self-induced human catastrophe or, if not, it will certainly be finished off sooner or later by some astronomical disaster, life might have existed elsewhere but is now extinct.

We are looking for very faint traces of something in an unimaginably, though not incalculably, vast space. We are looking for an amazingly small needle in an extraordinarily big haystack.

If we live on the only planet that ever had life, or one of the few, the chances that it happened were microscopically small and we are – to say the least – lucky. This has made me think not only about the big question of whether there is life in the Cosmos, but also, even more significantly, what the answer might imply.

My head tells me one thing about the possibility that there is intelligent life elsewhere in the Cosmos. My heart suggests another. I can be hopeful, more as a matter of belief than of astronomical knowledge. Although there were obviously some gigantic flukes in the way our planet evolved to create an environment that was benign to the development of intelligent life here, life is a versatile phenomenon and evolution seeks out the available routes to development. Life finds its way, and, once it has taken hold, it maintains the environment that it considers favourable. If this is indeed the case, there are reasons of universality – the same lines of thinking that guided the Greek atomists – to expect that, if and when we come face to face with sentient extraterrestrials, whether on their land or ours, they will look a bit like us, and have the ability and the desire to communicate. Both of us might be territorial, imperial and seeking space in which to expand, but I hope that curiosity, interest and benevolence will enable each to live and let live. I hope we enjoy and learn from the commonalities and differences between us, and that we both continue to appreciate the energy of life.

Notes

Introduction (pages 7–11)

1 'Billion' is not a formal scientific term, since it is ambiguous: in British English a billion is 1 million million, and in American English it is 1 thousand million. So I try to avoid using the word. But this can be cumbersome, and if I use it I follow the usual convention of informal scientific language that a billion is 1 thousand million.

Chapter 1 (pages 12–15)

1 Genesis 1:26.

Chapter 2 (pages 16–26)

1 In astronomical usage, and in this book, the word 'Earth', capitalized, means our planet. A lower-case 'earth' is a rocky planet similar to our Earth; a gaseous planet, such as Jupiter, is the archetype of other similar planets called 'jupiters'. The word 'earth' also means the substance of which the ground is made, or the Aristotelian element, as formerly understood. Likewise, the word 'Sun' means the star at the centre of our Solar System, but 'sun' means any star that has a planetary system. 'Universe' and our 'Solar System', as proper nouns, the names of unique objects, are also capitalized in astronomy, for consistency's sake.

2 Some bright stars have individual proper names, such as Vega. Others are named after the constellation in which they are located, with a Greek letter or a numeral, such as 51 Pegasi, which is in the constellation Pegasus. If a star has one or more companions, a Roman letter A is added to its name, with the companions called B, C, etc. The letter is a capital letter if the companion is a star, or a lower-case letter if it is a planet, as with 51 Pegasi b, the first planet found in orbit around the star 51 Pegasi A. Many stars are so faint that they are known only by their number in a catalogue, usually defined by the name of its individual compiler or team of compilers.

3 This cumbersome name can be decoded as 'the first planet discovered that orbits star number 7 in a catalogue of exoplanets discovered by the space satellite COROT'.

4 Likewise, this is 'the first planet discovered that orbits star number 10 in a catalogue of exoplanets discovered by the space satellite Kepler'.

5 This is a star in the catalogue, compiled principally by the early twentieth-century German astronomer Wilhelm Gliese, of stars that are closer than 25 parsecs (about 80 light years) to the Sun. Including some stars that might, through some error of measurement, really lie just over the 25 parsec boundary of the sample, there are about 5,200 of them in the latest revision of the catalogue.

Chapter 3 (pages 27–49)

1 This simplification assumes there is no immigration or emigration, and that there are no changes in the average lifetime, for example due to medical advances.

2 The Cosmic Microwave Background, a Nobel Prize-winning discovery in 1964 by American radio astronomers Arno Penzias (b. 1933) and Robert Wilson (b. 1936), is a background of radio noise that comes from the whole sky. It forms an irreducible background of confusion to every radio transmission. It is a cosmic radio signal that was generated in the hot material of the Big Bang and has persisted even to the present day.

3 Although Drake never heard an intelligent signal from either of the two stars, he did discover the (then secret) U2 spy plane. Slewing the telescope across the sky from one star to the other, he detected a strong, narrow-band, pulsed signal coming from somewhere overhead. His first reaction was that he had succeeded in his goal to detect an extraterrestrial radio signal. 'It can't be as easy as that,' he thought. It wasn't. The source of the signal was traversing the sky at a rate that indicated it was flying at a height of 80,000 feet, and was either a flying saucer, or an aircraft flying higher than the capability of any known aeroplane. Drake took his time to mull over the possibilities. CIA pilot Francis Gary Powers's U2 aircraft was shot down over the Soviet Union the following month; the existence and properties of the U2 became public, and it was clear to Drake what his signal had been. Drake's search was in the minds of Cambridge radio astronomers in 1967, when graduate student Jocelyn Bell (b. 1943) and her supervisor Antony Hewish (b. 1924) discovered celestial radio sources called 'pulsars', an acronym based on the description of their radio signals: 'pulsating radio stars'. Thinking that it was possible that the first examples they discovered could be some sort of interstellar radio navigation beacon, the radio astronomers only half-jokingly began to refer to the radio stars as LGM 1, LGM 2, etc., where LGM stood for 'Little Green Men'. But the signals were broadband, so they were not the sort of signal a radio engineer would make, and their regular beat showed no sign of a Doppler shift due to orbital motion in a planetary system. Pulsars were soon shown to

be natural phenomena, and the LGM designation was dropped for the more neutral PSR, for 'pulsar'.

4 Charles Messier (1730-1817) was an eighteenth-century French astronomer who specialized in searching for comets. He made a catalogue of items that he and others had found that might be confused with comets, including nebulae, star clusters and galaxies. Messier 13 is the thirteenth entry in this catalogue.

5 That 'yet' shows where my emotions lie. One day I hope we will detect that first signal, although I fear we will not (see chapter 16).

6 Attempting to emulate Kepler one day, I made myself into an international laughing stock. My friend Tim Radford, the then science correspondent of *The Guardian*, telephoned me one Sunday while I was still in bed reading the papers. On Sundays, the science correspondents are looking for light-hearted stories for tomorrow's newspapers that will lighten everybody's Monday morning. Tim was seeking comment on the discovery of a planet that had a very large force of gravity. What kind of life might it have? 'Gravity is strong,' I said, relaxed in bed, off my guard, 'so creatures would have to be robust. You, Tim, living as you do on Earth, are tall and gracile. Creatures on this new planet would be short and fat.' On Monday I found myself quoted in *The Guardian* under the headline, based on the words of Mr Spock to his captain in television's science-fiction series *Star Trek*, 'It's Life, Jim, But It's Short and Fat'. For several days, I watched with mounting chagrin as the foolishness that I had mouthed propagated on the World Wide Web throughout the world's science press.

Chapter 4 (pages 50–66)

1 Changing the chirality of an amino acid means that it has a completely different effect. The drug thalidomide is a chiral molecule; one form of it is a sedative, very safe to use. Manufactured in experimental quantities by natural processes, the correct form is made. But, built up by industrial processes from basic components, a thalidomide mixture is racemic, with both left- and right-handed forms produced equally. Used in the 1950s and 1960s in the racemic form for medical treatment, the presence of the other-handed form of thalidomide produced terrible side effects in pregnant women, including grave birth defects for tens of thousands of children.

Chapter 5 (pages 67–75)

1 A 'craton' is an old and inactive area of rock embedded in a tectonic plate; in contrast to more recent and more active areas of rock, which are often near the edges of the continents.

2 A 'terrane' is a fragment of the Earth's crust that has broken off from one tectonic plate and attached itself to another.

Chapter 7 (pages 82–94)

1 Meteorites found in Antarctica, and some other locations where post offices are scarce, are named differently. They are given a code designation, such as EET79001. This is the general location where they were found, the year of discovery and a serial number. The Allan Hills location in Antarctica is abbreviated as ALH, Elephant Moraine is EET, Queen Alexandra Range is QUE, and Meteorite Hills is MET. EET79001 was the first meteorite from Elephant Moraine that was numbered in 1979.

2 These technical terms describe three particular geometric figures. A planet orbits around its sun in a closed, elliptical orbit. A comet, dropping in towards its sun from a great distance, has an orbit that is open-ended, in the shape of a parabola. An interstellar comet, plunging in towards a sun at speed, has an orbit that is wide open, in the shape of a hyperbola. It was Isaac Newton who showed that all orbits under the force of gravity were members of this family of shapes, known mathematically as 'conic sections' – because if you have a wooden cone and slice through it with plane cuts of a saw at different angles, the exposed shapes are each one of the three geometric figures.

3 An 'autoclave' is a tank of high-pressure steam used to sterilize medical equipment etc.

4 Pronounced 'Vilt-2'.

Chapter 8 (pages 95–110)

1 The International Geosphere-Biosphere Programme (IGBP), the International Human Dimensions Programme on Global Environmental Change (IHDP), the World Climate Research Programme (WCRP) and the international biodiversity programme DIVERSITAS.

2 In Greek mythology Theia was a Titan (one of the earliest deities of the Greek pantheon), who gave birth to a daughter, Selene, the Moon goddess.

3 A gyre is a spiral motion, like a giant whirlpool.

4 Light is a wave, with the colours of the spectrum being waves with different wavelengths. The wavelength of light is small, and is measured in nanometres (nm). One nanometre is a billionth of a metre, one thousandth of the diameter of a human hair.

5 Chert is a hard sedimentary rock that forms in the sea from the tiny shells of siliceous sea organisms. Flint and jasper are forms of chert.

Chapter 9 (pages 111–22)
1 Mitochondria, hydrogenosomes and chloroplasts are specialized areas within eukaryotic cells that are responsible for respiration and energy production.

Chapter 10 (pages 123–32)
1 *Steppenwolf* is German for coyote, literally, 'wolf of the steppe', where the steppe is a large area of flat unforested grassland in Russia or Siberia. It is also the title of a book by Hermann Hesse (1877–1962), an autobiographical novel about his loneliness and animalistic nature.

Chapter 11 (pages 133–41)
1 Water with the heavier oxygen isotope is not 'heavy water', used in nuclear fusion experiments; heavy water is a water molecule with a heavier hydrogen isotope. The heavier hydrogen isotope is far less common than the heavier oxygen isotope, so it can be ignored in this discussion.

Chapter 13 (pages 174–92)
1 Strictly speaking 'geology' refers to the study of rocks on the Earth ('ge' is Greek for Earth) but 'areology' has the more general meaning of the study of everything about Mars. Surely the science of the rocks of both Earth and Mars is similar and part of the same study, so 'geology' in my book includes in its field of study the rocks of Earth, Mars and any rocky planet.
2 'Thermokarst': a land surface pitted by the thawing of subsurface permafrost. (The word was coined by analogy with the word 'karst', which, in geology, is the appearance of a landscape like the pitted limestone landscape of the Karst plateau region of Slovenia). 'Pingo': a mound of earth-covered ice (a West Canadian Inuit word). 'Polygonal ice wedges': polygonal forms in the ground made by repeated melting and freezing of ice. 'Esker': a long, sinuous ridge of sand and gravel, formed by the deposit of sediment on the bed of a stream running in an ice-tunnel within or under a glacier; when the

glacier melts, the streambed is lowered on to the underlying solid surface. 'Drumlin': a hill over which a glacier has flowed, giving it a characteristic hump-backed shape, like a whale's back, aligned with its neighbours in the direction of flow. The word drumlin comes from the Gaelic word *droimnín* ('little ridge'). 'Hoddy-doddy' for the same thing is a derogatory British-English dialect word for someone who is obese with a big bottom, transferred to glaciology because of the shape of drumlins. 'Glacial moraine': unconsolidated soil and rocks transported by a glacier and dumped from it in an accumulation.
3 1 nanometre = 1 millionth of a millimetre = 1 thousandth of a micron.

Chapter 14 (pages 193–99)
1 'Extremophilophiles', I suppose we may call them, or 'philoextremophiles'.

Chapter 15 (pages 200–5)
1 Planetologist John Zarnecki of the Open University in the UK, one of the principal investigators for the Huygens lander, has made a tongue-in-cheek suggestion that, like Enceladus, Titan could also be a future tourist destination. With its high, slow waves, 'Titan is the best place in the Solar System for surfing.'

Chapter 16 (pages 206–12)
1 According to legend, one of the physicists present in the canteen at the time responded 'We are right here and we call ourselves Hungarians.' The quip refers to the knowledge of nuclear physics that had been developed with the help of such physicists as Leó Szilárd (1898–1964), as well as to the uniqueness of the Hungarian language and culture.
2 Here I am setting aside the unverified testimony of individuals who claim to have seen strange craft in the sky, to have participated in autopsies of small humanoids from crashed spacecraft, or to have been subjected to internal medical examinations on board a flying saucer. I will believe such stories when they have been backed up by important and credible scientific evidence available for public verification.

Glossary

amino acids Organic molecules essential for terrestrial life because they build up into chains to make proteins.

archaea Primitive, single-celled organisms without cell nuclei.

arthropod An invertebrate animal with an external skeleton, segmented body and jointed appendages, such as legs. Insects, crustaceans, centipedes and spiders are arthropods.

asteroid A rocky, minor *planet*, of a size up to 1,000 km or so, orbiting in the Solar System, typically in the *Main Asteroid Belt* between Mars and Jupiter and in the *Kuiper Belt* beyond Neptune. Asteroids may have originated as planets with arrested development, broken pieces of planets that collided, or extinct comets, and some may have been captured by planets and become moons (or satellites). A very few are large enough to be called *dwarf planets*. Small asteroids, less than about 1 metre in size, are called *meteoroids*.

atomism The theory that the physical universe consists of minute, indivisible particles clustered together.

Big Splash The theory that the Moon was formed from mantle material scattered into space when a *planetesimal* called Theia struck the Earth in an early stage of its development.

biopoesis The process by which living matter might evolve from inanimate matter.

biosphere The entirety of the part of the *planet* that contains life.

black hole An astronomical body that is both small and massive, thus exerting such a strong force of gravity that no light or other radiation can leave the surface.

black smoker A *hydrothermal vent* under the sea, from which flow steam and other gases.

Cambrian explosion The stage in the evolution of life on Earth 542 million years ago marked by a dramatic increase in the number of living organisms.

catastrophism The theory that geological features are caused by unpredictable, large-scale events (catastrophes), such as floods or meteor impacts. Compare to *gradualism* or *uniformitarianism*.

chaos A physical effect, the property of certain complex systems (such as the weather or planetary positions), the future development of which is so sensitive that even the slightest change in the initial conditions would produce a completely different outcome. Prediction of the future of such a system is impossible beyond some initial period.

chirality Asymmetry of the type seen in the human hand in which the mirror image of an object cannot be superimposed on the object itself, no matter how the image is rotated.

chondrite A type of *meteorite* made of 'chondrules' (near-spherical globules) fused together in a matrix material, but the material of which has not been melted, as has happened, for example, in the interior of a *planet*.

comet A small body in orbit in the Solar System; a comet resembles an *asteroid* but is made of icy material, which, melted by the heat of the Sun, becomes gaseous and lets loose a cloud of solid dust particles, surrounding the nucleus with a coma and producing a dusty tail.

corona The hot atmosphere of the Sun (and other stars), extending from the visible surface of the star into interplanetary space.

coronal mass ejection An explosive event in which a cloud of solar material is ejected from the Sun's corona into space.

cyanobacteria Single-celled organisms, blue-green in colour.

degenerate matter According to quantum mechanics, there is a limit on how close two electrons can approach each other, which gives rise to a very strong pressure as you try to compress materials containing a high proportion of electrons. Degenerate matter is matter in such a state; it is found in nature in *white dwarf* stars. The pressure of the degenerate matter in such stars is what keeps them up, acting against the force of gravity that tends to compress them.

Doppler shift The change in wavelength or frequency of a moving source of waves – seen, for example, as the sound waves from a police car's horn or train's whistle change in pitch as the vehicle passes by.

dwarf planet A small *planet* or large *asteroid* orbiting the Sun, not one of the eight major planets, and not a satellite, large enough to have become spherical under its own gravity, but not so large that, as it formed, it cleared its neighbourhood of other minor planets and *planetesimals*. Also, an *exoplanet* of the same type.

ecliptic The plane of the orbit of the Earth, and by extension the line that this plane makes when it intersects the celestial sphere (the imaginary sphere centred on Earth).

eukaryote An organism composed of cells that contain a nucleus.

exoplanet A *planet* in a planetary system outside the Solar System.

extremophile A creature that thrives in extreme environments.

gamma-rays The most energetic form of electro-magnetic radiation.

gradualism The theory that geological formations are created incrementally.

gravitational micro-lensing A sudden brightening in the appearance of a star, caused by the magnifying effect of the gravitational field of a star or *planet* that passes across our line of sight.

gravity mapping A geological technique in which the variation in gravitational force on the Earth's surface is measured and mapped to reveal underground formations.

habitable zone The range of orbits that position a *planet* at just the right distance from its sun to have a life-friendly temperature. Also referred to as the Goldilocks zone.

heliopause The boundary of the heliosphere, the spherical region around the Sun that is filled with solar magnetic fields and the solar wind.

hydrothermal vent A fissure in rock that ejects water heated by geothermal sources below the ground.

Kuiper Belt The belt of *asteroids* and/or *comets*, or other trans-Neptunian objects (objects that orbit beyond Neptune).

Late Heavy Bombardment In the early history of the Solar System, the short period in which a rain of *meteorites*, *asteroids* and *comets* crashed down on the surface of the *planets*.

lichen A plant formed from the symbiotic association of certain fungi and algae or *cyanobacteria*.

magnetite A naturally occurring iron mineral akin to rust, with a crystal structure that generates a magnetic field; a 'lodestone'. Particles of magnetite are also produced by some living species, including some bees, birds and bacteria, and enable them to detect the magnetic field of the Earth.

Main Asteroid Belt The part of the Solar System, between Mars and Jupiter, in which orbit many *asteroids*.

meteor, meteorite, meteoroid A meteor is the name given to the phenomenon of a streak of light, radar echoes etc. caused by a small, solid body dropping from space into the atmosphere of Earth or another planet. The solid body is called a meteoroid and could equally be called a small *asteroid* (one less than about 1 metre in diameter). If any part of the meteoroid survives its passage through the atmosphere, it is called a meteorite.

Milankovitch cycles Oscillations in the Earth's temperature caused by variations in the planet's orbital parameters (its eccentric orbit, tilt, etc.).

nebula A body of gaseous material and/or dust grains in space that emits or reflects light and other energy picked up and redirected from stars nearby; the material from which stars are formed.

neutron star A star so small that its constituent material is made primarily of neutrons, as opposed to electrons and protons or other nuclei.

nucleotide One of the individual components of one string of the double helix molecule, DNA; the primary nucleotides of DNA are called adenine, cytosine, guanine and thymine (often abbreviated to A, C, G and T). Arranged in a particular order, they constitute the genetic code for a particular terrestrial species.

obliquity angle The angle (currently about 23.5 degrees) between the Earth's equator and the plane of the Earth's orbit; thus, the tilt of the Earth's polar axis relative to its orbit.

Oort Cloud The hypothetical region on the periphery of the Solar System from which *comets* come.

panspermia hypothesis The idea that life might be transferred from one planet to another.

photosynthesis The process by which plants convert sunlight into food. Carbon dioxide is converted into oxygen as a by-product.

phylogenetics A science that deploys the sequences of genes in organisms to infer the evolutionary relationships between species.

planet One of the eight bodies in orbit around the Sun, so large that it is spherical (or near spherical, if it rotates), having grown by clearing out the material in nearby orbits, its 'feeding zone'. Also, a body like this in another planetary system. Contrast *dwarf planet*.

planetesimal In a newly formed planetary system, a small solid body, like an *asteroid*, formed from accreted dust, which can potentially merge with others to form a larger *planet*.

prebiotic chemicals Chemicals which, although they are not biological, could be used to develop life.

precession The gyrating motion of a *planet* (notably the Earth) as its rotational axis describes a cone in space; precession is the gyrating motion of a spinning top.

prokaryote An organism, usually single-celled, the cells of which lack a nucleus.

pulsar Short term for 'pulsating radio star': a rotating *neutron star*, showing regular and rapid pulses of radio waves.

pyrolize To break down organic matter by subjecting it to high temperatures, without the presence of oxygen.

racemic A description of a material containing equal numbers of molecules that are left-handed and right-handed. See *chirality*.

radiometric dating Estimating the age of rocks by measuring the decay of certain naturally occurring radioactive isotopes.

refractory material Solid material that can survive high temperatures; early in the history of the Solar System, such dust particles accumulated, in the hot zones near the Sun, into the terrestrial *planets*.

resonance An arrangement of two *planets* in which an exact number of orbits of the one planet equals a different exact number of orbits of the other.

SETI Search for Extra-Terrestrial Intelligence – a label given to a number of research projects, beginning in 1960, that have sought to detect alien intelligence.

siderophile A chemical element that has a deep affinity with iron and is thus found in abundance deep in Earth's iron-rich core. Iridium and gold are both siderophiles.

Snowball Earth A hypothetical period (possibly one of several) during which the whole of the Earth's surface was covered by ice and glaciers.

snowline Figuratively, the boundary in the Solar System, between Mars and Jupiter, that separates the cooler outer reaches of the Solar System, where the gases in the *nebula* that created the *planets* remained solid, as ices, from the warmer region closer to the Sun.

'splosh' crater A type of crater formed by *meteorite* impact into icy mud, so that the mud flows out into a distinctive flower-like shape around the crater. Such craters are found on Mars and suggest that a layer of ice lies beneath the surface of the planet.

steppenwolf planet Free-floating, cold planets, not in orbit in any planetary system.

stromatolite Layered rocks in a stack like pancakes, formed by the growth of blue-green algae and other micro-organisms, which trap small grains of material (such as bits of sand or limestone) and bind them into a succession of horizontal mats, making a column that sticks up from the sea floor.

sunspot A disturbance in the magnetic field of the Sun that manifests itself as a dark blemish on its surface.

supernova A major stellar explosion that disrupts a star and releases a prodigious amount of energy, including a bright burst of light.

Thomists Medieval philosophers, following the lead of St Thomas Aquinas, who sought to combine the ancient thought of Aristotle with Christian beliefs.

trilobites A group of crab-like marine creatures (now extinct) that lived in the seas and on seabeds for 270 million years.

uniformitarianism In geology, the theory that geological formations are created slowly and incrementally; the geological equivalent of *gradualism*, as opposed to *catastrophism*.

white dwarf A small hot star that supports itself by the pressure of degenerate matter in its core.

zodiacal light A cone of sunlight, reflected by *meteoroids* and *asteroid* dust, that shines up from the horizon after sunset.

Further Reading

Crowe, Michael J., *The Extraterrestrial Life Debate 1750–1900*, Cambridge, 1986

Davies, Paul, *The Eerie Silence: Are We Alone in the Universe?*, London, 2010

Dick, Steven J., *Plurality of Worlds: The Origins of the Extraterrestrial Life Debate from Democritus to Kant*, Cambridge, 1982

Dominik, Martin and Zarnecki, John C. (eds), 'The Detection of Extra-terrestrial Life and the Consequences for Science and Society', *Philosophical Transactions of the Royal Society*, 369, 497–699, 2011

Kasting, James, *How To Find a Habitable Planet*, Princeton, 2010

Nield, Ted, *Incoming! Or, Why We Should Stop Worrying and Learn To Love the Meteorite*, London, 2011

Sagan, Carl, and Shklovskii, I. S., *Intelligent Life in the Universe*, San Francisco and London, 1966

Shuch, H. Paul (ed.), *Searching for Extraterrestrial Intelligence*, Berlin and Heidelberg, 2011

Sources of Quotations

Pages 12–13: Based on H. G. Wells, *The War of the Worlds*, London, 1898.

Page 14: Kathryn Denning, 'Is Life What We Make of It?', *Philosophical Transactions of the Royal Society*, 369 (2011), 669–78. Quotation from pp. 672–73.

Page 15: Genesis 1: 26.

Page 16: Blaise Pascal, *Pensées and Other Writings*, tr. Honor Levi, Oxford, 1995, p. 73.

Page 30: Nikola Tesla, 'Talking with the Planets', *Collier's Weekly*, 19 February 1901, 4–5.

Page 31: 'Radio to Stars, Marconi's Hope', *New York Times*, 20 January 1919.

Page 31: Editorial, *New York Times*, 21 January 1919.

Page 37: Carl Sagan, 'The Lifetimes of Technical Civilisations', in Carl Sagan (ed.), *Communication with Extraterrestrial Intelligence*, Cambridge, MA, 1975, p. 147.

Page 37: Paul Davies, 'Searching for a Shadow Biosphere on Earth as a Test of the Cosmic Imperative', *Philosophical Transactions of the Royal Society*, 369 (2011), 624–32. Quotation from p. 626.

Page 43: Paul Davies, *The Eerie Silence: Are We Alone in the Universe?*, London, 2010.

Page 44: *Epicurus: The Extant Remains*, ed. and tr. Cyril Baily, Oxford, 1926, p. 25.

Page 44: *On the Heavens by Aristotle*, tr. W. K. C. Guthrie, Cambridge, MA, 1939, Book I, Ch. 8, 276b, lines 10–12.

Page 47: Andrew D. Weiner, 'Expelling the Beast: Bruno's Adventures in England', *Modern Philology* 78 (1980), 1–13.

Page 47: 'Il sommario del processo di Giordano Bruno, con appendice di documenti sull'eresia e l'inquisizione a Modena nel secolo XVI', ed. Angelo Mercati, in *Studi e Testi*, vol. 101.

Page 47: Steven J. Dick, *Plurality of Worlds: The Origins of the Extraterrestrial Life Debate from Democritus to Kant*, Cambridge, 1982, pp. 71–72.

Page 50: Kevin W. Kelley (ed.), *The Home Planet*, New York, 1988.

Page 63: Birger Schmitz quoted in Ted Nield, *Incoming! Or, Why We Should Stop Worrying and Learn to Love the Meteorite*, London, 2011, p. 197.

Page 83: Charles Darwin, letter of 29 March 1863, repr. in *The Correspondence of Charles Darwin*, vol. 11, ed. Frederick Burkhardt *et al.*, Cambridge, 1999, pp. 277–78. Quotation from p. 278.

Page 89: T. C. Chamberlin and R. T. Chamberlin, 'Earlier Terrestrial Conditions that May Have Favoured Organic Synthesis', *Science*, 28 (1908), 897–910. Quotation from p. 900.

Pages 89–90: Charles Darwin, letter of 1 February 1871, repr. in *The Correspondence of Charles Darwin*, vol. 19, ed. Frederick Burkhardt *et al.*, Cambridge, 2012.

Pages 90, 91: Stanley Miller quoted in Sean Henahan, 'From Primordial Soup to the Prebiotic Beach', October 1996 (http://accessexcellence.org/wn/nm/miller.php).

Page 94: 'British Association, Meeting at Dundee, Origin and Nature of Life', *The Times*, 5 September 1912, p. 5.

Pages 95–96: Amsterdam Declaration on Global Change, 2001 (http://www.essp.org/index.php?id=41).

Page 103: Henri Poincaré, *Science and Method*, tr. Francis Maitland, London, 1914, repr. 2003, p. 68.

Page 116: Simon Conway-Morris, 'Predicting what Extra-Terrestrials Will Be Like: And Preparing for the Worst', *Philosophical Transactions of the Royal Society*, 369 (2011), 555–71. Quotation from p. 556.

Page 119: J. D. Watson and F. H. C. Crick, 'A Structure for Deoxyribose Nucleic Acid', *Nature*, 25 April 1953, 737–38. Quotation from p. 737.

Page 128: Pete Conrad quoted in Alfred Crosby, 'Micro-organisms and Extraterrestrial Travel', in *Humans in Outer Space: Interdisciplinary Odysseys*, ed. Luca Codignola *et al.*, Vienna and New York, 2009, pp. 6–13. Quotation from p. 9.

Page 140: Gerta Keller quoted in Chris Hedges, 'Professor Chips Away at Theory on Dinosaurs' Demise/Scientist Tackled Adversity on Road to New Hypothesis', *Houston Chronicle*, 1 April 2004, p. 18.

Page 144: Luis W. Alvarez *et al.*, 'Extraterrestrial Cause for the Cretaceous-Tertiary Extinction', *Science*, 208 (1980), 1095–1108. Quotation is from the article's abstract.

Page 151: Georges Cuvier, *Essay on the Theory of the Earth*, New York, 1818, p. 38.

Page 151: Sir Charles Lyell, *Principles of Geology: Being an Attempt to Explain the Former Changes of the Earth's*

Surface by Reference to Causes Now in Operation, vol. 3, London, 1833, p. 6.

Page 151: Archibald Geikie, *The Founders of Geology*, New York, 1905, p. 299.

Pages 174–75: H. G. Wells, *The War of the Worlds*, London, 1898.

Page 176: William Herschel, 'On the Remarkable Appearances at the Polar Regions of the Planet Mars, the Inclination of its Axis, the Position of its Poles, and its Spheroidical Figure: With a Few Hints Relating to its Real Diameter and Atmosphere' (1784), repr. in *William Herschel: Collected Scientific Papers*, ed. J. L. E. Dreyer (London, 1912), vol. I, pp. 131–56. Quotation from p. 156.

Page 177: E. M. Antoniadi, 'Recent Observations of Mars', *Knowledge*, 25 April 1902, 81–84.

Page 188: 'Life But No Bodies on Mars', *New Scientist*, 14 October 1976, p. 78.

Page 190: Statement by President Clinton released by the Office of the Press Secretary, 7 August 1996.

Page 191: David S. McKay *et al.*, 'Search for Past Life on Mars: Possible Relic Biogenic Activity in Martian Meteorite ALH84001', *Science*, 16 August 1996, pp. 273, 924–30.

Page 193: Stillman Drake, 'Galileo's First Telescopic Observations', *Journal for the History of Astronomy*, 7 (1976), 153–68. Quotations from pp. 157 and 161.

Page 194: Edward Rosen, ed. and tr., *Kepler's 'Conversation' with Galileo's 'Sidereal Messenger'*, New York, 1965, p. 42.

Page 203: Carolyn Porco quoted in Alan Boyle, 'Liquid Water on Saturn Moon Could Support Life', NBC News, 9 March 2006 (http://www.msnbc.msn.com/id/11736311/ns/technology_and_science-space/t/liquid-water-saturn-moon-could-support-life/#.UAVr18owIVI).

Page 210: G. D. Brin, 'The Great Silence: The Controversy Concerning Extraterrestrial Intelligent Life', *Quarterly Journal of the Royal Astronomical Society*, 24(3), 1983, 283–309.

Page 210: David Darling, *Life Everywhere: The Maverick Science of Astrobiology* (New York, 2001), p. 103.

Page 211: P. R. Lawson and W. A. Traub (eds.), 'Earth-like Exoplanets: The Science of NASA's Navigator Program', report published by NASA on 16 October 2006, p. 10 (http://exep.jpl.nasa.gov/files/exep/NavigatorScience2006.pdf).

Illustration Credits

1   ESO/A.-M. Lagrange
2   NASA/ESA and L. Ricci (ESO)
3   ESO/Sergey Stepanenko
4   NASA/ESA and L. Ricci (ESO)
5   ESO/H.H. Heyer
6   Ohio State University Radio Observatory and the North American AstroPhysical Observatory (NAAPO)
7   NASA
8   Arecibo Observatory
9   Camille Flammarion, *L'Atmosphere: Météorologie Populaire*, Paris, 1888, p. 163
10 NASA/JPL-Caltech
11 ESO/S. Deiries
12 NASA/JPL-Caltech/UCLA/MPS/DLR/IDA
13 NASA/JPL-Caltech/DLR
14 ESA/SMART-1/Space-X (Space Exploration Institute), ESA/SMART-1/AMIE Camera Team
15 Paul Murdin
16 NASA/SPL
17 Frans Lanting/Corbis
18 Eye of Science/SPL
19 ESA/DLR/FU Berlin (G. Neukum)
20 NASA, ESA, J. Hester and A. Loll (Arizona State University)
21 Yohkoh mission, ISAS/JAXA, Japan, with NASA and SERC/PPARC (UK). Yohkoh Legacy Archive at Montana State University, supported by NASA
22 NASA/Goddard/SDO AIA Team
23 Walter Alvarez/SPL
24 Virgil L. Sharpton, University of Alaska, Fairbanks. Lunar and Planetary Institute
25 NASA and The Hubble Heritage Team (STScI/AURA)
26 NASA/JPL-Caltech/Arizona State University, R. Luk (vertical exaggeration 2.5x)
27 NASA/JPL-Caltech/Cornell
28 NASA/JPL-Caltech/University of Arizona
29 NASA/JPL-Caltech/ASU
30 NASA/JPL-Caltech/University of Arizona/Texas A&M University
31 NASA
32 NASA
33 NASA/JPL-Caltech/University of Arizona
34 NASA/JPL-Caltech/GSFC/SWRI/SSI
35 NASA/JPL-Caltech/Space Science Institute
36 NASA/JPL-Caltech/USGS

Index